Anatomy of Life

Born in Kolkata, Devdan Chaudhuri was educated in India (Fergusson College, Pune) and England (University of Essex). While pursuing his Masters, he lost the inspiration to become an economist and returned to Kolkata to work on his writings and research. Over the years he became an entrepreneur in the art and hospitality sectors. He enjoys travelling and photography. *Anatomy of Life*, his debut novel, was nominated for the Tibor Jones South Asia Prize 2013.

Anatomy of Life

Devdan Chaudhuri

PICADOR INDIA

First published in the Indian subcontinent 2014 by Picador India
an imprint of Pan Macmillan India, a division of
Macmillan Publishers India Limited
Pan Macmillan India, 707, Kailash Building
26, K.G. Marg, New Delhi – 110 001
www.panmacmillan.co.in

Pan Macmillan, 20 New Wharf Road, London N1 9RR
Basingstoke and Oxford
Associated companies throughout the world
www.panmacmillan.com

ISBN 978-93-82616-30-6
Copyright ©Devdan Chaudhuri 2014

This is a work of fiction. All characters, locations and events
are fictitious. Any resemblance to actual events or locales
or persons, living or dead, is entirely coincidental.

All rights reserved. No part of this publication may be reproduced, stored in
or introduced into a retrieval system, or transmitted, in any form, or by
any means (electronic, mechanical, photocopying, recording or otherwise)
without the prior written permission of the publisher. Any person who
does any unauthorized act in relation to this publication may be
liable to criminal prosecution and civil claims for damages.

1 3 5 7 9 8 6 4 2

This book is sold subject to the condition that it shall not, by way of trade
or otherwise, be lent, re-sold, hired out, or otherwise circulated without
the publisher's prior consent in any form of binding or cover other than
that in which it is published and without a similar condition including
this condition being imposed on the subsequent purchaser.

Typeset by Glyph Graphics Private Limited, Delhi 110 096
Printed and bound in India by Gopsons Papers Ltd

For the youth

For the youth

Contents

One

Seasons   3

Two

Myriad Void   89
Circles and Spheres   143

Three

Centre and Periphery   167
Balance   203
The Wheel   227

ONE

Seasons

1

After the divorce of his parents, when the poet and his mother travelled to another city to begin a new life, the poet's mother often used the despondent phrase 'everything is finished'. The end of a marriage, separation from one's firstborn and shifting to a new city to start all over again aren't the ideal circumstances for a woman in her forties. The poet knew that his mother's 'emotional hangover' was justified, for it is a difficult task to detach oneself inwardly from deep attachments formed over many years. And moreover, after a period of shock, hurt and bewilderment, she had landed in a new city devoid of the simple assurance of the known and the familiar.

She had to get used to the new – the office, the people, the streets, the shops, the markets. And unlike him, his mother didn't meet the unfamiliar like an enthusiastic explorer, eager for discoveries, but like an uncertain child, far away from home, in the midst of the alien and the unknown.

But in time, the poet told himself, his mother would get used to her new life; everything would ease away – the unknown would become familiar, and the exile would turn into a home.

1.1

Despite his spirit of optimism, the melancholic phrase 'everything is finished' caused within him an irritation and a persistent unease. For the first few instances, he met the statement with silence, hoping that his mother would finally let it go, and prefer not to infect the present with the bitterness of the past. As months passed, the poet's mother continued to be in a state of mind that dulled her awareness and slowed her thoughts. A silent numbness caused by bitter reflection continued to haunt her. One evening at the dinner table when his mother once again uttered the tragic statement with a heavy sigh, something burst within the poet. 'Why do you always say everything is finished?' he questioned. There was anger, earnestness and understanding in his voice. 'Why don't you look at our life in a different way? At least, all those anxieties are gone. Shouting. Screaming. Madness. We are free of them. You have a good job. We have this apartment. Mother, it's not an end, it's a new beginning.'

He was silent for a moment and then said, 'Life cannot remain the same throughout. You know that. Phases begin and end. Seasons of time continue to change. We cannot stop them. But we can live through them in the best possible manner. Please mother, don't utter that phrase ever again. You will achieve nothing by spoiling your mood.'

The poet himself was surprised at his own words. He never knew they existed within him. It was the first time that the poet had encountered the phrase 'seasons of time'.

1.2

The poet was eighteen when he passed his higher secondary examination with a distinction in economics. He had been living with his mother for two years and the opportunity to travel to a new city to pursue a degree in economics spread within him an excitement of freedom and adventure.

His mother was also pleased that her son had found a place in one of the most reputed Arts and Science colleges of the country.

'Hostel life will also do you a load of good,' she had told him while he was packing his luggage. 'A new life. Another new beginning.'

2

Four large stone buildings occupied the north of the campus. The poet's room was on the first floor of the third block. The room was modestly furnished with an iron cot, a wooden desk, a chair and wall shelves. The only luxury was the large south-facing window; it had a splendid view of lush green trees, the basketball courts, the dispensary, the dusty field, the roof of the library and the amphitheatre.

The door to the room opened out to a stone passage, which connected all the rooms, washrooms and staircases. The passage overlooked a square with concrete benches, adolescent trees and a lamp post in the middle.

The room had a wooden ceiling. Some of the previous occupants had made an effort to reach the ceiling and write their names and the years of occupancy. All the names were old-fashioned. They had to be – the years of occupancy pointed to pre-Independence days.

The college was established in 1885. The Gothic-style buildings spread themselves within a large campus marked with trees, pathways and gardens.

The two playing fields in the west lead to the hillocks which were frequented by fitness freaks and lovers.

3

The poet began his college life with great fervour. He joined the student body, became a leading member of the quiz team and an active coordinator of the literary and film societies. He was on good terms with professors and college seniors and this added to his reputation. He came to be known as one of those who could get things done, be it organizing events or winning prizes on behalf of the college.

The poet was popular and had a large circle of friends, among whom one was the pianist.

4

The pianist had become impatient with the poet. Two months had elapsed. They had met in the campus on the first day of college. Then they had visited pubs, bookshops and restaurants, but beyond that nothing had happened.

She had made her desire for the poet obvious, it was all written in her gaze, her smiles, her double entendres, her casual reference to the fact that she had the apartment all to herself during the day. The poet had noticed the signs, but had chosen not to respond. And that ensured the poet's presence in her thoughts. She was forced to offer him the curious vigilance that women reserve for men whom they fail to understand to their satisfaction – not enough to either love or reject.

Then she thought of a new plan. She would play a game. The student body had organized a fresher's party. The poet would be there, so would that guy from her class, who was smitten with her.

5

At the party, the pianist searched for the poet. Someone told her that the poet was jamming with the college rock band in the recreation hall. The informer also mentioned that the poet's mobile was switched off, but, he added, the poet remembered the party and would come after a while. The pianist looked at her watch. It was half past ten.

The party was on the terrace of a budget hotel. She walked down the stairs to the toilet. A couple of drunk guys were banging on the bolted door. Then the door opened. A half-naked girl was sitting on the toilet seat. Her red dress had fallen down to her hips. Her red bra was on the floor. She seemed drunk. There was a guy buttoning up his jeans. The guy walked out; the two drunkards went in and bolted the door.

The pianist found herself another toilet. She kept thinking about the midnight deadline at her home. She felt angry at the poet for being late. She stared at the mirror, calmed herself and went back to the terrace.

The music had slowed down and the lights had dimmed, couples could now kiss and fondle each other in the dark. She walked to a corner and waited for the poet as she surveyed the dance floor and the terrace door.

The pianist's admirer had been stalking her for an hour. Finally, he gathered some courage, walked up to her and asked her for a dance. But the pianist paid no attention and turned him down.

A moment later, she glimpsed the poet at the door to the terrace. 'Hey, wait,' she called out softly to her admirer. She led him to the periphery of the dance floor, put her arms around his neck, made sure that her breasts were pressed against his chest, and started to sway in slow deliberate movements.

At that very moment a ruckus erupted. A few guys had got into a fight. A volley of loud and distinct swearing made the fight look dead serious. The shrieks of frightened girls filled the locality. The music stopped. Bright lights ruined the atmosphere.

The pianist saw the poet, along with a few others, trying to protect a guy from being beaten up by three angry assailants. Flashes of light reflected from a small knife that hadn't been used yet. 'I will tear you to pieces,' shouted the most overzealous assailant, who was being forcibly curtailed by the peacemakers.

The stupid fight finally stopped when the three assailants had been pinned to the ground. The victim, whose left eyebrow had split, was also bleeding through his nose. He had made the mistake of dating the ex-girlfriend of one of the assailants.

The hotel manager arrived in a bad mood and the party was over. The poet stayed back with the group to try and pacify the manager, who threatened to complain to the police and wanted names. Everyone else was asked to leave.

In life, things are not going to be according to one's liking – a simple fact that is often the hardest to accept. The pianist felt the truth of this fact. Her plan had failed due to the unforeseen incident.

The pianist sighed and made her way out of the terrace. She followed the crowd of chattering students down the staircase.

6

When the pianist arrived at college on Monday morning, a girl from her class informed her that the poet was looking for her. This had never happened before. She immediately forgot the promises she had made to herself (not to waste time and energy on the poet) and found the poet in the café that faced the college's main entrance.

The pianist noticed that the poet was looking different. His mood was playful and his words bold.

The pianist quickly caught on with the poet. When he gulped an analgesic tablet calling it his magic pill, she said, 'A magic pill for a woman means something else.'

The poet smiled and lit a cigarette. Then he said, 'Two things I find very attractive about you. Your dazzling smile and your talent as a pianist.'

'I have other talents as well,' the pianist remarked.

Both of them looked at each other. Their eyes met. A certain gravity deepened the moment of brazen silence.

'Why don't we go to your apartment? You can play some Beethoven for me,' the poet suggested, already knowing what her reply would be.

7

The poet and the pianist reached the modest building in thirty minutes. Its location was somewhat secluded. 'My father has bought another flat. We will be shifting to the new one after a few months,' the pianist explained on the staircase. Then she turned to the poet and smiled, 'But we will still keep this one. A few things will stay here. And the keys will always remain with me.'

They came to a heavily protected door on the second floor. The pianist searched her bag for the keys. The poet's heartbeat, his deepening breath and the feeling at the pit of his stomach left him in no doubt of what he was anticipating.

In the living room, the poet was struck by the sparse décor. It was almost Japanese. The living room had the mighty presence of a grand piano, a couple of abstract paintings of human figures, an artificial plant, a few chairs and a couch.

The poet wanted to take a closer look at the grand piano. But the pianist held him from behind and pulled out his shirt. The poet turned to meet the eager lips of the pianist. After a while, without any hesitancy and with a cool confidence, they undressed like they were used to each other for years.

The poet could sense that the pianist was undressing with the aim of enticing him, to make him a bit impatient. She wanted the poet to desire her, impatience delighted her.

For the next couple of weeks, excluding the weekends, the poet spent all his afternoons in the pianist's apartment.

8

Relationships are often tested not only by the act of lovemaking but also by what happens immediately after the act.

After three months from the day the poet had first visited the pianist's home, he realized that his lover captivated him, but only till his climax, after which a strange emptiness surfaced within him. He didn't feel any need to remain with the pianist. His mind swayed to other things and his body displayed signs of restlessness. He hurried away citing excuses, and avoided spending time with the pianist anywhere else other than the apartment.

The poet made no effort to understand his discomfort to become free of it. His initial impulse was to run away from the unease by carrying it deep within him. He occupied himself with other things, and tried to avoid the pianist, as if the pianist had suddenly become burdensome to him.

But the pianist hadn't become a burden, the poet had become a burden to himself. By avoiding the pianist, he sought to avoid the discomfort that had surfaced within him.

He failed to relieve himself of the weight. It resurfaced within him when the student activities got over, when the movies he watched ended, when the classes he attended terminated, when he woke up from sleep.

When the pianist found the poet in the campus, she was perplexed by the sudden change in his behaviour. He appeared melancholic and no longer enticed her with his smiles and laughter. The poet spoke to her about more solemn things; the free-flowing attitude of the poet had become dull.

The poet made excuses for missing out on their dates. 'I have a few things to follow up. It's about the film festival. Got to go to the archives. We will be able to meet only after a week.'

But the very next day something inside him relented. He once again found himself in the pianist's bedroom.

The pianist noticed the waning of the poet's enthusiasm, and reacted by heightening the poet's pleasure, that began to reach its zenith with unusual frequency.

'You like to watch, don't you?' the pianist observed after another kiss. Her mouth smelt of the poet's groin. 'You always keep your eyes open while kissing. I read in a book that a woman shouldn't trust a man who kisses with his eyes wide open,' the pianist said.

'Don't you trust me?' the poet asked instinctively and immediately felt an unease at his own question.

'I surely do,' the pianist replied, almost lovingly, and climbed over the poet.

Her words rekindled the sensation of unease and discomfort; he became aware of a deep heaviness that spread around his throat and his chest.

9

Within a week of their meeting, it was obvious to the poet that the pianist fancied him. But he wasn't entirely sure whether to go beyond the casual friendship that had formed between them. He decided to give himself more time to be absolutely sure of his understanding of her.

That night at the party he had panicked. When he had noticed the pianist dancing with that guy from her class, he had felt a surge of jealousy in the form of an anxiety – the fear of losing the pianist. Provoked by the fear, he had rushed into the affair. And immediately after that, the fear had vanished, and desire had taken over.

10

Before the beginning of the affair, the pianist had displayed a passion for his concerns and agreed to his views about life. But afterwards, the poet realized that she was only interested in the poet, not in his concerns, and by doing so she separated herself from the vital part of the poet which governed his sense of self.

No two people are alike. But a relationship tends to weaken when the people involved fail to understand and respect each other's passions and beliefs.

In other words, every person feels alive with thoughts which are important to them. And when someone else cannot relate to those thoughts, then it becomes impossible to relate deeply with him or her.

Without those deeper connections, relationships, which are meant to be intimate, become shallow and superficial.

The pianist wasn't what the poet had thought her to be. All the serious words they had been exchanging all this while actually meant very little to the pianist. Unlike the poet, she wasn't looking for deeper insights about life. The hours spent together discussing earnest ideas were nothing but a game to impress and seduce each other. This fact had disappointed the poet. His deepest self could not be engaged with her.

11

The melancholy of discovering an aspect of a person that is quite different from one's previous understanding of the same person is one of the saddest human experiences.

But the wave of unease, melancholy and discomfort that was troubling the poet did not stem from that understanding. Neither did it arrive, when the poet's infatuation – formed of desire, deceptive understanding and panic – began to weaken.

Even when the relationship – devoid of any emotional intensity – gradually became mere acts of sexuality, the poet could adjust within himself, and allow desire and excitement to propel his attraction for the pianist. He knew that he didn't love the pianist and that his relationship with the pianist would not go deeper than an erotic friendship. (He also realized that in a relationship, one has to relate to the mind and the soul; sexual relation alone means nothing.) With this understanding he was relieved of the obligation to involve his soul. He went solely with the impulse of his body, and felt free to engage himself with his lover, whose passionate performance was alluring and captivating.

Till then it was alright.

When the soul is uninvolved it's one thing, but when the soul starts to trigger a niggle of discontent, then it becomes a different matter altogether.

He had begun to sense that the pianist didn't look upon him in the same manner he looked upon her. The unease stemmed from his sense of guilt. He was suffering from the guilt of viewing the pianist solely with desire, while she viewed him with affection.

12

A few afternoons later, when the drops of sweat which had accumulated on the poet's forehead broke off to trail down the pianist's neck, the pianist imploringly cried out, 'I love you very much. Oh! How much I love you.'

Later that night, the poet encountered a manifestation of a crises that he wasn't familiar with, a bitterness seeped into the taste of his cigarette, a monstrous burden possessed his chest and a moral dilemma surfaced out of himself, and got him to reflect.

On one hand was the temptation of pleasure, and on the other hand, the niggle in his soul, that manifested itself as an uneasy guilt for viewing the pianist only with desire, while the pianist probably loved him and had reposed in him all her trust.

The poet didn't feel any attachment for the pianist. If she vanished from his life it would make no difference to him. But at the same time, he couldn't live with this truth about himself, the fact that his sole motivation to spend time with the pianist was triggered by mere eroticism. He felt like a wicked manipulator trying to equate love with desire. He wanted to do what was fair – terminate the relationship, and bid the pianist a graceful farewell.

On the other hand, he thought it would be an act of foolishness to forfeit the pleasures and deny himself on obscure grounds. There was no doubt that he desired the pianist. So it was in his interest to keep alive the relationship and forget all about his stupid moral qualms.

After a while some minor aversions passed his mind. He was expected to avoid the notice of the neighbours and slip into the apartment like a thief, a role he didn't appreciate. He wasn't allowed to smoke in the apartment because of the sensitive nose of the pianist's mother; and the scare he got one afternoon when the doorbell rang – he had to hide in the pianist's bathroom, engulfed in anxiety and shame. He had felt the greatest relief of his life when the pianist told him it was just a door-to-door salesperson.

Despite such things which impeded freedom, the lure of touch was still strong.

Images of their intimacy pranced within his mind and aroused temptation, while the feeling of guilt, originating from his soul, protested furiously, by churning out looms of anguish.

It was an incredible sensation for the poet – two sides of his own self competed fiercely to grab his awareness, his decision, his response. He continued to suffer. The temptation of pleasure and the surge of conscience vied to conquer him.

The poet lit another cigarette, went out of his room, stood in the passage and looked at the courtyard, and then at the dark sky, beyond the western roof of the hostel.

The sight of the stars could have comforted him. But his awareness lay within, for that is where the battle was raging. On the surface, his face was blank, contemplative and unhappy.

After a while, something happened – his mind was able to free itself and usher in a moment of calm.

He noticed the cool breeze that brushed against his face and the moon that spread angular shadows over the square of the courtyard.

And at that very moment he achieved clarity.

13

'Please don't ask me to give you reasons,' the poet said, 'I won't be able to see you anymore.'

The pianist remained silent. She looked more irritated than upset. Her brisk breathing announced the arrival of anger. Her reaction relieved the poet who had feared melodrama. The pianist's look was a far cry from that of a tragedienne. 'It's your decision,' she said. The poet didn't prolong the heaviness that had descended upon the living room; he left the pianist after giving a final kiss on her cheek. Everything ended as gracefully as it had begun.

The poet closed the door behind him and climbed down the staircase in a hurry.

When he came out to the street and caught a glimpse of the afternoon sky, an incredible feeling erupted within him. A sense of elevation hurled him into an intense moment of weightlessness.

From the dark chasm of his dilemma, he suddenly experienced the vast sky of a weightless mind – the most important characteristic of the feeling that is judged as happiness.

14

The poet returned to the hostel with an enthusiasm that surprised his friend, the joker. The poet somehow looked different. He was brimming with buoyancy.

'You seem to be very happy,' the joker remarked.

'I have discovered a great truth,' the poet replied.

'Do we need to celebrate then?' the joker asked.

'We should. Let's roll some joints,' the poet proposed.

Fifteen minutes later, while smoking a joint, the joker finally asked, 'So, what great truth did you find out?'

'There are two forces in every human being,' the poet began, 'One comes from the ego and the other comes from the soul. These two forces compete between them, and try to grab hold of your response. When this conflict occurs, your character depends upon your response. Whom do you allow yourself to be governed by, the will of the selfish ego or the will of the soul.'

The poet took a deep puff, slowly exhaled and then continued, 'Conflict between them creates the moral sense or the conscience. And when you act in tandem with your conscience, something wonderful tends to happen. You feel happy and light. Your self-esteem and confidence are boosted.'

'Or in simple words,' the poet said, 'one who doesn't violate the conscience finds happiness.'

15

The poet had discovered something that everyone ultimately discovers – the conscience or the moral sense, the niggle within one's soul, has the power to create happiness.

The ability to defeat one's self-interest (in the poet's case, sexual pleasure) by following the instinct of one's moral sense, tends to raise confidence and produces a feeling of satisfaction, buoyancy and lightness.

It may seem that the poet denied himself the pleasure in order to satiate his sense of morality. But on that occasion the poet didn't curb his desire, he gave vent to it. His desire to fulfil his innate moral sense was greater than his desire to violate it.

Morality doesn't require self-denial. There had been no displeasing self-denial, but a happy self-fulfilment.

16

It is vital to understand the difference between moral values and moral sense.

A couple of generations ago, if a young woman was known to possess a habit of going to the bioscope to watch a film, then her chances of finding a husband from a respectable family actually diminished.

Watching movies is no longer considered immoral, and likewise, killing someone as an act of self-defence is now permitted and justified. Societies keep changing their perception of what is moral and immoral. The evolution of the moral perception is a continuous process, where the moral values are laid down, broken and reconstructed, amidst constant debate, friction and disagreement between different mindsets.

Different mindsets create their own values. One may have conservative or liberal values, social or capitalist conditioning, left or right inclinations, but all those values have nothing to do so with morality which is the natural inheritance of a human being.

Morality in its essence is neither conservative nor liberal, neither socialist nor capitalist, neither middle class nor upper class, neither literate nor illiterate, nor left, right or centre. Morality is the presence of conscience, a sense of righteousness or the moral sense. It is a natural tool of the human self, that recognizes and values *fairness*.

To be immoral is to be unfair, to be moral is to be fair – this is the natural moral dictum of humans.

17

When the poet and the pianist met in the campus, neither of them stopped for any conversation.

But the pianist did what a woman does best – use the speech of facial expressions to convey her message. Whenever their eyes met, her face gave out a stern unfeeling look that carried a whiff of anger.

The look indicated something more to the poet. He realized he had overestimated his effect on her. He hadn't touched her soul or caused her grief. If he had done so, then the manner in which the pianist looked at him would have been very different. It would perhaps have been a look of sombreness, reflection and even sadness.

Perhaps it was a mistake to assume that the pianist had begun to view him with affection. Perhaps they were on the same side, viewing each other as desire dolls, partners of pleasure, sexual objects.

The thought made him smile. But he didn't regret his decision. A relationship without love is neither satisfying nor engaging. When there is no soul, there is no sense. He did what he had to do. And he had no regrets.

The pianist gave a performance in the college auditorium that won her many admirers.

The poet soon learnt that the pianist had a new lover. He felt a peculiar sense of relief – oblivion must have been granted to him by her memory.

He stopped thinking about her, without any effort.

18

The poet only slept at around 2 a.m. and that negated the possibility of attending early morning lectures.

After he woke up, his routine activity was a pleasant monotony that he cherished. Earlier he used to switch on his alarm clock, but he soon realized that if he woke up with the alarm, the chances of remembering his dreams became slim. He slept very deeply and liked to regain his waking consciousness very slowly. He maintained his sleeping posture till the awareness of his environment became acute and distinct. Then he got up from bed, drank some water, smoked a cigarette and surveyed the newspaper that was slipped under the door by the newspaper man. He didn't read the newspaper thoroughly. He marked the items of interest in his memory, and waited till he returned in the afternoon to read the paper in detail, before it was borrowed by someone or the other, and never returned.

After twenty minutes, the poet got dressed and reached the fountain near the main building.

The poet had to interact with many other students. He met them as usual. Fifteen minutes later, he walked out of the main entrance of the college, crossed the street and sat down at his usual table at the café.

The university examinations were only a month away, it was time again for resolutions, hard work and black coffee. Everyone was busy collecting the notes they had missed. The reading halls were frequented by the serious students, and those who thought it was the best opportunity to find themselves a lover.

The poet had also picked up the pace in his academic pursuits. He was assisted by a friend who ran a voluntary tutorial in his room to help those who had been negligent.

The poet usually ordered south Indian breakfast while he sipped tea and smoked a couple of cigarettes.

Once, while gazing at the busy street, the main gate of the college, the trees and the sky, he noticed the day moon. He wondered at her subtlety and detachment. He acquired a metaphor – as detached and subtle as a day moon.

Then he thought that the moon had only taken on the colours of his emotions – the way he felt at that moment – a little removed from life around him.

Recently, something had been changing within him – his awareness of himself was becoming deeper and even his desires were being modified and refined. He struggled to keep alive his fun-filled interests – fun was not fun anymore – and sought something more substantial.

So he gave more time to his habit of reflection – to sense, to feel and to think.

Every night, he would read something, reflect over it, make notes and listen to music.

Somehow he was more aware than before, and the desire to fill himself with the deeper ingredients of his own self, had taken a firm root.

19

The sweetheart was the poet's acquaintance, who for the past one year had sought the poet's attention. She had followed the poet in the campus, had attended all the student activities where the poet had been involved and had collected the newspaper clippings of the youth-related articles which the poet had begun to write, for the Saturday supplement of a local English daily.

Four months into the second year of college, the poet fell in love with her.

The sweetheart carried the appearance of a certain innocence – symbolic of a beautiful goodness – whose attraction had prompted the poet to delve into spiritual philosophies and develop a love for them. It was this state of his mind, dominated by the soulful, that enabled him to notice the girl whom he had been acquainted with for over a year. The poet was drawn towards her as he had been drawn towards his quest for wisdom, and the sweetheart responded with a shy happiness that contributed wholesomely to her beauty.

She also inspired the poet's poetry:

I could feel your love
in your searching eyes,
bent towards the earth
like flowers in the rain,
struggling to look up
and confess.

I could feel your soul
within the moon of your smiles.
With the breeze of your glance
I could feel your touch.

I could feel your hope,
a simple hope,
to be cared for,
and loved.

20

The poet and the sweetheart got along beautifully, and delighted themselves with the nuances and gestures of love. Simple things suddenly meant more, and they lived in happiness; love had banished everything else from their minds.

They frequented cinemas, restaurants, pubs, night clubs and shopping streets. Soon they got tired of crowded places and began to favour the seclusion of long walks, gardens and parks, where the sense of togetherness magnified.

21

The sweetheart had a lady friend who had given her the second set of keys to her apartment that was only visited on Sundays for the weekly cleaning. Since the sweetheart's father came home very late in the night, she could easily manage to give an excuse to her mother and spend time with the poet in the apartment that was quite close to the college campus.

It was already nearing ten. The night outside had settled in with a calm serenity. The poet was feeling the strange intoxicating drowsiness that occurs when feelings of love and passion collide and flow together like water at the confluence of two rivers.

It was a dazed sensation that he only felt after he had been with the sweetheart. It was a mingled sensation of fulfilment, of love and of desire, of happiness and pleasure.

The poet felt drowsy and wanted to fall asleep, with the sweetheart by his side.

The sweetheart rested her head on the poet's chest. Her hair smelt of shampoo. 'I love the fragrance of your skin,' she whispered.

For a woman, the natural odour of a man can be one of the ingredients that determine attraction or repulsion.

'You smell like something sweet,' she added. The poet laughed softly. 'My grandmother says the smell of the body reflects the odour of the soul. You can know a person by his smell.' 'Yes,' the sweetheart whispered, 'I like the fragrance of your soul,' and laughed like a child.

The poet could take advantage of his natural odour and skip baths without nauseating anyone in close proximity. But now he made it a point to clean himself well – the sweetheart didn't like the bitter taste when she moved her tongue within his ears, sometimes biting gently. She asked the poet for love bites on parts of her body which were always concealed with clothes.

The poet put his arms around the sweetheart and began to caress her skin. Only in a relationship of love the gentle caresses of touch gain more importance. The poet had already realized: before a man learns to make love to a woman, he must learn how to touch a woman.

'You know,' the sweetheart confessed mischievously, 'I am addicted to your kisses.' She smiled, lifted her head and sought the poet's lips.

They lost themselves in what mattered more in love, much more than sexuality, the prolonged act of the kiss.

They pecked each other like two love birds, and then, the moist movements became more vigorous and deep.

Freed of everything, they felt and sensed the spacious purity of love, and the astonishing miracle of a fleeting deliverance.

Later that night, the poet wrote:

> Love breaks forth as a kiss –
> nothing else can release love,
> the unforgettable dissolution
> of lips pleading love, as guilty.

22

The sweetheart came home thirty minutes before midnight. The house was at the dead end of a deep street lit brightly by white fluorescence.

As she approached the house, she felt relieved to not see her father's car in the driveway or inside the garage.

Her father stayed at home only from midnight till 10 a.m. – she was free to do as she pleased. Without switching on the garage lights, she parked her sleek scooter, put a blue plastic cover over it, dusted her hands against her jeans, avoided spotting the terrifying cockroaches scurrying across the garage floor and hurried out.

She walked across the small lawn and entered the house through the back door that the maid always kept unlocked as per her instructions. She had to pass through the kitchen and the dining space to arrive at the living room. Her dog had heard her coming and greeted her with jubilance.

All dogs have their own character, but they often take on the personality of their masters or mistresses. 'You know, my dog,' the sweetheart would begin, while talking to her friends in the gym, 'she is also an Aquarius like me. She stares blankly out of the window, likes to keep to herself and doesn't like being pushed around.'

She always referred to the animal as a dog even though the dog was a female, a bitch. (The sweetheart's father had only allowed his young daughter to keep a bitch for a pet.)

'No one waits for me like you do,' she said to her dog as if she were speaking to a child. She cuddled her dog and felt the joyous vigour of a moist happy tongue licking her face.

She closed the door of her room and went about her business quietly. Her mother wasn't asleep. She could hear the melodramatic sounds of a movie channel – only a five-inch wall separated her room from her parent's bedroom.

Muted years of dull marriage had turned the sweetheart's mother into something unresponsive and voiceless. The sweetheart's father was the dominant male in the house. Every household decision, ranging from what kind of food would be served to the guests to the choice of a kitchen gadget, had to be authorized by her father. Since her childhood, the sweetheart was taught to view her father from a distance. Her father was someone who wasn't openly loved or hated. Everyone had to do what he pleased. No one had the audacity or the courage to challenge him. Whenever her father was around there was a sense of fear and uneasiness – jokes were cut short, voices became low, everyone shot off in different directions.

It was also expected that her father would terrorize and mistreat everyone. No one bore the brunt of his temper more than the sweetheart's mother. But she never rebelled. She silently endured the insults and the occasional slaps, which made her ears ring, and imprinted red bruises on her sensitive face.

The people who lack consideration make truths sound like something cruel and unfair. The sweetheart's father called his wife a social parasite – someone who eats on the resources and contributes nothing. He revelled in making incisive remarks that wounded those parts of her which his slaps couldn't reach.

As a defence mechanism, the sweetheart's mother had acquired the ability to become unresponsive like a lifeless mass of stone. She escaped into the world of television that stirred the emotional responses which her own life could never evoke – her husband behaved like a brute and her grown-up daughter had many other things to do than spend time with her. (Deep down she also believed that both her husband and her daughter thought her to be stupid, completely unaware of the changing realities of the world – someone unable to converse intelligently.)

Television took care of her isolation. She got addicted to her favourite channels, but the harmless diversion soon turned into a chronic addiction. 'My wife suffers from TV – the television virus has infected her,' the sweetheart's father often joked to his friends.

But the ailment was more serious. In times of a cable failure, she kept staring at the screeching white noise and behaved anxiously as if something terrible had happened. Every month, she tipped the cable guy to keep her connection trouble-free and wrote his number in four diaries, and kept one of them inside a locker. She only lived within the twists and turns of the daily soaps. She didn't have any friends, didn't go anywhere except the odd visit to a relative or a market, and from morning till late night, she watched television, pausing three times every six hours to ask the maid to give her the prescribed eye drops.

The sweetheart wasn't completely insensitive to her mother's problems. 'When you completed an MA in history,' the sweetheart would ask, 'why didn't you do something, other than sitting at home? You could have taught at least; you could have had your own life.'

'It was a condition of the marriage that I should only take care of the home and the family. Your sister was born within a year of our marriage. There were things to do. Your grandmother used to live with us. There used to be something of a family life. And to speak frankly, I had no desire to do anything else. There was no need.'

'But it's never too late,' the sweetheart said. 'You can get involved in some work that might make you happy.'

'Who said I am unhappy!' she retorted. She looked at the cuckoo clock, went to her room, and switched on a serial.

From that day the sweetheart had stopped hoping. She knew nothing would change at home. She feared her father, and her mother didn't have the ability to understand her viewpoint. Home was just a house for her. There was no one to rely on, fall back upon or talk to. She lacked the guidance of someone older, who would understand her. This somehow made her very vulnerable and very insecure.

Things had taken a turn for the worse when the sweetheart's elder sister died of a rare illness whose medical name was so long that it was impossible for her to remember. She was only ten years old then, but she had been deeply affected by the morbid gloom that had settled over the house ever since the disease had been diagnosed. The treatment had lasted a year, operations were conducted, but everything failed.

The morbid phase in their house never gave way to a brighter one. The season of gloom did clear, but it left behind a grave hollowness. There was no sense of family, of togetherness – everyone lived within their own orb of loneliness.

The vacant atmosphere was dominated by the fear generated by her father's presence, the noise of the television and a cold uneasy feeling of a deserted ruin. Her father was busy with his work and his circle of friends, her mother was a captive of the television, and she only longed to escape the cold graveyard of home, once and for all, and as soon as possible.

23

During the third year of college, the poet and the sweetheart were coming to terms with the fact that they were two separate individuals, with their own distinct natures, preferences, beliefs, priorities, likes and dislikes.

The critical phase where they had to accept each other as they were, instead of trying to transform each other, had begun.

Love was there as before – the urge to kiss was intense – but the meaning that they sought to thrust on love was different. The sweetheart started asking questions, 'Once you had told me when one has fallen in love it is not always important that he should live with the person one loves…' The poet contemplated. 'I know I had, but I don't feel that way any longer. I used to think if I feel love then that should be sufficient. I used to think differently, now it's something else.'

Then she asked him the question that every man has to confront. 'Do you love me?' 'Yes,' said the poet. Then she asked him why. The poet laughed and said he didn't know. But the sweetheart was determined to know the 'reasons'.

'I need to know. Tell me,' she pressed. As if the 'reasons' were her fundamental need, as basic as food, water and sleep.

The poet never tried to use flattery to conquer the minds of women, to force them gently to think of him, when they recalled the secret pleasure that the flattery had caused. He wasn't interested in the kind of women who would fall for flattery. The sweetheart was immune to flattery – she responded to it with suspicion. But all women love praises, and the sweetheart was no different.

There is a fine difference between flattery and praise. Flattery raises the ego, while praise raises the soul. Flattery is to tell a woman – you are the most beautiful woman, you are very intelligent, you are better than her in every way and so on, while praise is to tell a woman about the *effect* she has on others. It is like saying: your presence makes me very calm, or, you trigger the goodness in me, and so on. Flattery is often an exaggeration, while praise is appreciation. Flattery can be based upon lies, but praises have to rely on truths.

The poet told her the truth – her rare qualities which he admired – she was neither demanding nor intrusive, she was reserved and graceful, she carried herself with a lady-like dignity, but the sublime restlessness of her soul reminded him of a playful butterfly.

Then he felt a wave, scribbled few lines on a sheet of paper and gave it to her. The poem delighted the sweetheart, and filled her with happiness.

> Love doesn't arise from the chessboard of reason,
> but from the spacious womb of reality within –
> deep as deep sleep, origin of my dreams,
> the sea of soul, that triggers this poetry.

24

Some days later, the sweetheart was flipping through the book that the head of the Department of German Language had given her to read. The poet didn't have the patience to learn German, he had dropped the optional subject after first year, but he had read a translation of the novel in English.

About the hero, who had shot himself above the right eye, blowing his brains out because of unrequited love, the poet had said, 'To kill oneself due to a rush of sentiments is so stupid. Only a fool would let his lover govern his entire sense of life.' The sweetheart disagreed and felt the need to say something. 'Perhaps nowadays one cannot love as intensely as before. So someone killing himself for the sake of love appears foolish,' she said. 'I don't think so,' replied the poet. 'No one would commit suicide for love. It only happens in sentimental books.'

The poet had spoken too soon. At midday, the tranquil murmur of the campus was shattered by two gun shots. The young man was still struggling when he was taken to the hospital, where he survived for a few painful hours, but the girl had died on the spot. The bullet had pierced her forehead.

A complete story finally appeared in the newspapers the next morning. The physics student was in a relationship with his classmate, but the girl had decided to end things after a heated altercation under the shade of the trees in front of the physics department. The young man left the girl to acquire a patriarchal FN 6.35 from his grandfather's closet. Then he walked towards a crowd of students who were having their lunch during the break. He singled out his lover and approached her. Then he pointed the pistol at her forehead and pressed the trigger. Then he shot himself. The recoil thrust made the bullet brush against his forehead. He convulsed for a night and died in the hospital.

It was rumoured amongst the students that the last words of the young man were inspired from the dialogues of popular movies. Before the young man shot the girl, he might not have said but had surely thought, 'If I cannot get you, then no one else will get you.'

The poet told the joker that love had nothing to do with the incident. 'Love can never destroy you, but a lover can.'

Then he pointed out that it was egoism, selfishness, hatred, attachment and a thirst for vengeance in the nature of the lover that had provoked the action, not the feeling of love.

Love had nothing to do with it, the poet had concluded with certainty.

The incident made the sweetheart gravely upset. She couldn't hold herself back when the joker made ghost jokes about it. 'How can you joke about something like this!' she protested. 'It could have been someone known to us. It could have been us.'

'Come on, I won't shoot you if you leave me,' the poet said. 'I know you won't,' the sweetheart said in a tone of irritation. 'Neither would you shoot yourself. You never take love seriously.'

The sudden accusation shocked the poet. The joker politely excused himself from the table. 'What's the matter with you?' asked the poet. The sweetheart was almost in tears. She picked up her sunglasses and put them on. A moment later, she stood up. 'True. A woman has to be a fool to love a man,' she said and walked off.

But she couldn't keep herself away for long. In the evening she called the poet and met him. 'I don't know what came over me. Please don't mind.' The poet said he didn't. 'The incident must have gone to your head.' The sweetheart agreed. 'Forget it. I don't want to think about it anymore,' she said. 'I have the keys to the flat. We have two hours. Let's go.'

25

Later that night while staring out of the window of her room, the sweetheart cursed herself for being so impulsive. She knew the poet hadn't deserved to hear what he did, that he didn't take love seriously. Now, it appeared absurd to her that she had actually expected that the poet would tell her. 'If you leave me I will also shoot you and then shoot myself.' She couldn't help but think that the words would have proved how much the poet wanted her.

She felt disappointed that the poet seldom displayed any signs of absolute burning love, love that one cannot do without, cannot breathe without. The poet didn't ever give her the opportunity to brush his hair and to console him during times of despair. The poet never sank his head into her lap and said, 'What will happen to me without you?'

The sweetheart always kept searching for words and gestures which manifested the poet's need for her. She loved the moments when the poet would ask her to run some errands for him. She felt needed, and that pleased her.

Once there was a minor problem. The poet hadn't received his admit card for the university examinations due to an error by the office clerk. He was tense, and the first thing he did was call her up and talk to her. At that moment, she had felt a deep happiness. Not because the poet was in distress, but because he had chosen her as the *support* for his distress. The problem was solved in a couple of days. But since then there had been no such anxiety that the poet sought to ease by confiding in her. And this made her feel a bit unhappy, a bit less wanted, a bit less important.

The sweetheart wanted to feel needed. And satiation of this deep-rooted need was paramount.

But somehow the poet didn't seem to realize that. And the sweetheart didn't have the confidence and the force to make him realize.

26

As time progressed the relationship began to pass through short phases during which the sweetheart felt removed from the poet. The poet would forget to give her a call, he would miss a date and pursue things which were more vital to him. And once they met, the poet would tell her things she really couldn't relate to, and whenever she would make an attempt to differ from his ideas the poet would still have the last word.

Sometimes she wanted to have the last word but she always failed in the objective, not because the poet gave her no opportunity, but because she quickly reconciled with the fact that she was incapable of halting the poet with any kind of understanding that the poet didn't possess or have a better version of. 'I know I am not as intelligent as you are,' she once told the poet with a contentment comparable to the pride of a person who feels delighted when his children display more worth than him.

But soon the matter of intelligence turned into a grievance. The poet rarely shared his darker moods with her. The poet had never felt the need to have someone to share his despairs with. He was habituated to falling upon himself as his only support – that was how he had prepared himself from his boyhood. He never spoke about his mother's death and his strained relationship with his father and brother. This attitude of the poet disappointed the sweetheart. She believed nothing could bring lovers closer than sharing intimate despairs. She surprised the poet by saying that he didn't think her intelligent enough to understand all his thoughts. The poet laughed and dismissed the accusation with a hint of charm, 'If you get to read all my thoughts, then you might be tempted to run me over with your scooty.'

The sweetheart's mind inferred a wild meaning from the poet's words.

That night the sweetheart dreamt that the poet was passing by in a car driven by some other woman. She had frantically waved to the poet but he had completely ignored her. Then she dreamt that the poet was making love to that woman on the backseat of the car – all the actions appeared vulgar and untamed. She had nearly panicked. She wanted to tell the poet about the horrible nightmare, but she couldn't meet the poet. He was busy with work, some event was being organized; he had no time to listen to her.

The sweetheart became aware of a stark truth. Unlike her, the life of the poet wasn't governed by his love for his lover.

After a lull – when the poet would simply forget about her – he would confuse the sweetheart with his outbursts of concern and affection.

The sweetheart soon realized that things would only continue in this way, a spell of detachment, and then, a sudden spurt of renewed intensity.

With this thought, a strange despair crept into her mind.

She thought about the poet and wanted to tell him:

> You cause me so many despairs
> that I have to confess, I love you.

27

A commitment had gradually grown between them. The poet realized it one afternoon, when he saw the sweetheart practising her signature using the poet's surname.

And one day when the sweetheart casually remarked that the poet would be a good father he felt an obvious uneasiness. The thought of marriage, kids and family didn't fill the poet with sentiments which the sweetheart loved to dream about. On the contrary, they made him a bit uneasy and reminded him of the endless quarrels and the violent scenes which he had witnessed as a boy. He had never thought about marriage and family; it was too early for him.

'A time comes when the woman has to know whether the man she loves intends to marry her,' the sweetheart told the poet a few days later in the apartment. 'Women have to prepare themselves, make adjustments, they can't just keep switching on and off.'

She said the words in context of a friend whose boyfriend confused her by promising marriage, and then kept on postponing it indefinitely. 'My friend says it has become very hard to find a guy who can commit nowadays.'

The poet could feel that the sweetheart was hinting at him, trying to elicit from him a comment about the future of their relationship. But he didn't say a word. He went to the window, drew aside the curtains, slid the glass and lit a cigarette. He could see the well-lit living room of the house next door. A portly woman was watching prime-time television, a plump girl of about ten years had spread herself on the floor with crayons and drawing books. The poet changed his pose and his angle of vision. He stared at the slice of sky that emerged beyond the roof of buildings with stars, clouds and a haze.

'Do you like dogs?' the sweetheart suddenly asked. The question was easy to tackle. 'Yes, I do, in fact,' the poet said and turned away from the window. 'My mother, at one time, was fond of dogs. She had an old cocker spaniel before I was born. I have seen its photograph. But after the dog died, my mother never raised any other.'

Then the poet told the sweetheart how, as a boy, he used to feed all the street dogs, and give them names and how they kept following him around...

'That's why I like dogs so much,' the sweetheart remarked. 'They practise the rare virtues of faithfulness, devotion and loyalty.'

28

Some hours later, in his room at the hostel, the poet sensed that he had failed to dedicate himself to the sweetheart in the manner she had done, and he often felt guilty about it – an inability that stalked the poet as his own shadow. Something in him always prevented him from saying 'anything for you' – a phrase that he had heard a hundred times from the sweetheart.

The poet knew, the sweetheart would do anything for him, even if he told her to do something that was unfair, illegal and immoral. This power of love frightened him; he viewed it as a weakness – a blindness – not a strength.

And he knew he could never say 'anything for you', in case that 'anything' wasn't agreeable to him and defied his sense of judgement.

But for the sweetheart, the only thing she wouldn't ever do was something *against* the poet. To cause him any harm would be unthinkable. And to go against the poet's will would mean going against her love – the force she felt within.

So whenever she said, 'anything for you', it also signified 'anything you say', 'anything you wish' – a deep-rooted longing to transform the force of love into a deed.

29

The sweetheart soon began to sense that her relationship with the poet may not materialize into marriage, but she had convinced herself, in the spirit of sentimentality, that she would keep on loving the poet, 'no matter what', even if the poet left her, or fell in love with someone else.

But she knew the poet would never betray her. She knew that there was something in the poet that appeared to her as noble. A quality that she missed in her own father. The poet was everything that her father wasn't. That was perhaps why she had been attracted to him in the first place. During her school days, whenever she was asked about the traits she would look for in her lover, she always replied, 'He has to have a sense of humour, he must be broad-minded, intelligent, well-behaved and sensitive.' The poet had fitted the mould – she had even thanked God for it. But she had always found him very different from all the young men she was acquainted with. He never told her what her friends' boyfriends told them – what colour bra she should wear, whom she shouldn't speak to, where she shouldn't go. The poet didn't seek to enforce his will upon her.

But often that very indifference caused her suffering. Sometimes when she would be in despair due to some problem at home she would seek the poet's solace. But the poet would tell her, 'Life will assault mercilessly if the mind is unprepared and the soul is weak. Trust me on this. Don't depend upon anyone else to solve your own despairs. Find solitude, confront yourself and gather strength.'

But for the sweetheart, solitude didn't mean the space to prepare oneself, solitude meant loneliness and boredom. At times such as these, she would feel a great surge of anger at the poet and would run away to cry upon her pillow.

But how well they laughed together! How well they kissed! How easy she felt with him! How well he touched her! How warm his skin felt! Her body was so cold. During winter when she would touch the poet he would nearly shiver. But he would clasp her cold hands in his warm ones and allow the heat and the cold to arrive at the tepid warmth of love.

Those little things made her happy, they were vital to her. The worries of the poet were alien to her. She liked her own little worries. How she thrived whenever the poet had fever! She would be so worried and so tense. She would take him to the doctor, buy medicines and fruits, cook soup for him and pray for him, as if he had some grave illness.

The poet, during his fever, would come out of his room and remain with her. He would listen to her like an obedient child. He would take all his medicines, he wouldn't have only cigarettes for breakfast, he wouldn't stay awake till very late.

But after recovery, he would say, 'Due to the bloody fever, one whole week was wasted.' And then again, he would disappear.

Such an attitude planted a shadow of doubt within the sweetheart's mind, regarding the poet's actual need for her.

But she also blamed herself for being unable to inspire the poet to greater intensity. Self-blame was one of her weaknesses.

30

The sweetheart didn't possess a habit of reading, she was far too busy being a yoga and Pilates instructor (earning her own money to buy gifts for the poet), attending extra classes of German, taking care of her dog, enduring the environment at home and loving the poet.

Earlier, when the poet wished to pass on to her the books which he deemed would be helpful for her to live and understand her life, the sweetheart used to say, 'You are there, you will know, that will be enough for us.' The poet didn't search for his own concerns in the sweetheart. He had realized that he was perfectly capable of loving a woman who didn't share his pursuits. It is the common values which keep people together, and the poet felt they had similar values, and that was enough.

But very soon the sweetheart started blaming herself for not being able to converse about the things which kindled the interest of the poet. She also couldn't understand why the poet had stopped writing love poems and wrote sad philosophical verses. She felt the poet wasn't happy in the relationship and blamed herself for it.

But that wasn't the truth. The poet was already suffering from a philosophical vacuum, and he tried to assuage it through his inquiries.

The poet wanted to learn of better ways to lead his own life. Examples set before him only taught him in reverse – what not to be, how not to act.

And he knew, for a start, there was no one to help him, except the library.

And more importantly, he was also going through the phase which every thinking person goes through – the need to stay unhappy, favouring the twilight of evening rather than that of dawn, withdrawing into oneself, disillusions of every kind, magnification of the life led in solitude than with people.

The poet began to avoid going to the cinema or to parties, and asked the sweetheart to go alone, or with her friends. Such moods of the poet confused the sweetheart. She continued to think that the poet was unhappy with her.

31

When the initial season of new love dies, another begins, where the individuality of each lover becomes prominent.

This is the critical phase of discoveries, acceptance and tolerance, when the weaknesses of one's lover tend to get magnified in the mind, instead of the strengths.

The poet began to spot many weaknesses in the sweetheart's nature. They were weaknesses from the poet's point of view. He accepted them – as a man accepts a silly wife or a woman accepts a brutish husband – but with a sense of disappointment, that diverted his attention to the other spheres of his life.

The sweetheart also allowed him to do what he wanted to, because she feared that if she encroached upon the poet's time, then he would feel disgusted and leave her.

But she didn't realize that the poet felt unhappy when the sweetheart never asserted herself. The poet wanted an equal relationship, where the sweetheart would demand her prerogative, but she felt more comfortable accepting the role of the lesser partner, the weaker sex.

Many other misunderstandings continued to occur. One cannot alleviate misunderstandings unless one talks about them, but their conversations had become irregular. What they didn't like about each other was kept buried, and invisible barriers cropped up each passing day.

And when honest conversations tend to breakdown, then the decline of any relationship becomes inevitable.

32

The sweetheart studied the German language and wanted to pursue advanced studies – a two-year course in one of the universities in Germany. One of her professors had told her about a scholarship that offered what the sweetheart wanted. 'You've got to take some extra lessons,' her professor had said. So she had enrolled herself in the successive levels of courses offered by the Max Mueller Institute. She had classes from morning till evening and met the poet only once in a while. She made new friends, one of them told her that he loved two women. She thought it was impossible to be in love with two people simultaneously. So she decided to ask the poet about it. 'Tell me, is it possible that someone who is in love, may feel something for another person?'

'Are you in love with someone else?'

'No,' protested the sweetheart, 'I am just asking. Can't I ask you anything nowadays?'

The poet remained silent. He felt very bored that the sweetheart couldn't talk about anything else other than love and its complex variations.

'Listen, please, for God's sake don't get wrong ideas,' implored the sweetheart. A couple of girls, occupying the next table, turned towards them and exchanged a humorous whisper. A smart alec on a flashy motorbike without a silencer sped past with a roar of noise. The poet didn't like the anxious look on the sweetheart's face. He smiled and spoke to her nicely. 'I am not getting any wrong ideas.'

The sweetheart felt relieved. She smiled and grabbed the hand of the poet and squeezed it with the firmness of her feelings. The poet withdrew his hand. He didn't want them to appear as teenaged lovers holding hands in a restaurant.

'I have to go,' the poet said. 'Got to do some work.'

'I thought we were going out for dinner tonight.'

'Not tonight.'

The poet felt disappointed that the sweetheart made no effort to make him change his mind.

The poet paid at the counter for the two cups of coffee and exchanged a few friendly words with the co-owner of the café. The co-owner asked the poet to wait and showed him the new photographs he had clicked.

The co-owner had a passion for photography. The poet had visited an exhibition where some of the photographs had been exhibited. All of them displayed the distress of street children. One of the photographs had also won a national level prize.

'Now that I have got the confidence,' the amateur photographer said, 'I will send a few of them abroad.'

'Perhaps you will win more prizes,' the poet remarked. The photographer smiled. 'I really hope so.'

33

The poet and the sweetheart crossed the road and walked to the parking lot within the campus, where the sweetheart's scooter was parked.

After the sweetheart sped away, the poet walked back to the same table they had been occupying and lit a cigarette.

He thought of the sweetheart's question and his response. Then he thought, how easily it was possible that someone else may replace him in the apartment of the sweetheart's lady friend. The idea discomforted him; he fell into a bad mood.

He gazed blankly at the busy road, the entrance to the college and the trees of the horticultural garden.

After he had drunk a cup of tea he felt better. Then he regretted turning down the dinner. They hadn't gone anywhere for weeks, and he really didn't have anything else to do. He let out a deep breath, called the waiter and asked for the bill.

34

The poet walked back to his room, sat on his chair and rested his feet on the edge of his table. The ticking of the table clock wasn't audible – the night stretched ahead.

He put on some music. His favourite rock band was Pink Floyd. He knew all the songs and had collected all the bootleg versions. The band members, once students of architecture, constructed poetics of space, depth and distance.

The poet loved the *sense* of the band – the expansive style of music was closer to classical symphonies. It offered him the harmonious shell within which he could feel, listen and introspect.

35

After an hour, the joker walked in and said he had the strangest of stories to narrate.

One of the guys, living in a private hostel, had found a girlfriend. The young man left every evening – with great confidence thanks to his smart new clothes – to spend time with the love of his life. He had told many stories which had made his friends envious. His lover was quite beautiful (7 on 10, which was the set standard of a popular actress). She came from an affluent but disintegrated family, liked to drink beer, visited a temple on Saturday mornings and a discotheque in the night. She found him handsome, charming and intelligent. The young man had promised his friends that he would introduce her to all of them.

A few weeks passed, and the young man continued to disappear every evening to meet his girlfriend. Every day he groomed himself; his friends were glad to see him wearing clean underwear and odourless socks, using an expensive deodorant – his abuses also lost their usual frequency.

Everyone started praising the power of a woman who had brought about such a positive change in him.

The young man used to be in high spirits and displayed the greeting cards which he received from her, like certificates.

Then, one evening, someone saw him at a faraway movie theatre when he was supposed to be with his girlfriend at her residence. When confronted, he simply collapsed. A little while later he confessed, in pathetic emotions, that he didn't have a girlfriend, and had been lying all along.

'Think about it,' joker said, 'he even bought a girlie handkerchief, sprinkled it with girlie perfume and showed it to his friends. Can you imagine? He wrote all those cards with his left hand. He starved himself to save money to buy presents for himself.'

The story made the poet feel ill, but the joker laughed and cracked jokes.

A long chunk of ash from the joker's cigarette fell on the bed and a tiny black stain remained on the bedsheet as the joker failed to clear it in time.

36

After the joker had left, the poet was overcome by tiredness – an emotional tiredness; a weariness of the mind, not body.

He needed silence. He sat on his chair and stared out of the window. The floodlit basketball courts in the distance caught his eye. The shrieks of the girls were blown in by a northbound breeze. The noise irritated him. He plugged in the earplugs of his MP3 player, lit a cigarette, went out of his room for his usual walk – slowly, quietly, in rhythm with his thoughts – in the dark stone corridor, overlooking the square.

Music never failed him when he needed it the most. But tonight he had been conquered by a monstrous feeling. He went back to his room, drank water, returned to the corridor, leaned over the wooden railings and stared down at the square. Then he looked up at the night sky and spotted the moon.

As a boy, while staring at the moon, the poet would blur and defocus his eyesight, to get a glimpse of the colours of the moon which flashed and danced around its periphery.

He did that once again – his vision blurred, but the colours came into focus. He could see the deep blue, outlining the silver disk, and the striking flashes of red.

After a while, he returned to his room, sat down to write something, and stared blankly at a white sheet.

An insight came to him – life only takes the meaning of the moment, be it a fleeting moment of happiness, or torment.

The poet felt the touch of something pure, like the joy of wisdom. He smiled. How true were the words: life can only take the colour of one's mind, one's feelings, one's thoughts. *Life is not a string of adjectives. I have suffered, it's not enough, I must seek and understand.*

37

The poet had first seen the virgin inside a lecture hall. He was held in attention by the virgin who intermittently stretched her chewing gum, holding one end between her teeth, and pulling the other end with the fingers of her left hand. She continued to stretch the sticky gum till it became so thin and so light that the strong breeze that was blowing in from the large French windows could make it flutter.

Before the thin white curve snapped, the virgin put it back inside her mouth, and continued to chew with a deliberate slowness. She went on to repeat her action, and the poet noticed that the virgin had an air of lazy sensuality about her.

After a few days, the poet had a vivid dream. He saw himself with the virgin in a deserted square surrounded by tall dilapidated buildings. It resembled a scene from a gangster movie – a dark neighbourhood of negligence and criminality. It was night. They walked to a dark building with a rusted fire escape and climbed the stairs to enter a second-floor room. The interior was as dark as outside. There was a bed in the centre of the room under an old-fashioned ceiling fan. The poet went to open a window while the virgin, sitting on the bed, undressed herself.

To be dreaming of making love to a woman who was a stranger in reality was an intriguing experience for the poet. The phrase 'dream lover' also took on a new meaning.

But the poet didn't try to view the dream as an omen or a prophecy; he didn't try to become a friend of the virgin. He dismissed the vivid scenes of the dream as one of the many absurd experiences which stamp around in the mysterious world of sleep.

Two years had passed since then. The 'dream lover' had remained just a casual acquaintance, but their friendship had matured quickly, during the final year of college.

38

One day the virgin walked up to the poet and said, 'I liked the passage that you wrote in the literary magazine. I also hold similar views.'

'That's good, I suppose,' the poet replied.

'If you are free we could have a chat over a cup of tea perhaps?'

'Yeah,' the poet consented. 'Every time is teatime for me.'

The café bore a deserted look. It was the time of day that the poet liked best. It was late afternoon, the blaze of the sun had weakened and the lunchtime crowd had dispersed. This was the best time to have a quiet conversation, before the crowd gathered once again in the evening.

'Writing is about discovery,' the poet was saying, 'even a simple diary can release you and make you aware of the things which are more true to yourself.'

'That's right,' the virgin agreed, 'But I never had the courage to keep a diary. I always feared someone else might read it.'

The waiter arrived with a glass of fresh lime juice and a cup of tea. The poet took a sip and called back the waiter – a new recruit. He told the waiter to change the tea. 'Get the sugar separately,' he instructed.

The conversation progressed and shifted to other topics. The virgin told the poet about her fear of failure – unless she was able to succeed to some extent, she felt that she would be ignored and no one would bother about her.

'But is it wise to think in this way?' the poet questioned. 'You should look at your life only through your own eyes, not through the eyes of others. You can't afford to live life like this. If you do so, you will end up being unhappy.'

The virgin remained silent. Then she asked, 'What is happiness?'

The poet thought for a while, lit a cigarette and then spoke. 'Usually one confuses well-being – prosperity and pleasure – with happiness. That's a mistake. Many economists have realized it as well. In one of the papers, they have tried to contrast the rich countries with the happy countries. According to them, the opposite of well-being is happiness.'

'Well-being is important,' the poet continued, 'but happiness is even more crucial. In one of my poems, I wrote, happiness is a deed, not an end. I believe that's the truth. One has to *do* happiness. And for that, one has to follow the will of the soul. That's the way to be happy. And to go against it, is a recipe for disaster.'

'But the meaning of happiness changes with time,' the virgin observed, 'What makes you happy at one point in time doesn't seem to satisfy you in another.'

'That happens because we change and evolve,' the poet replied. 'But one thing doesn't change, whether we are twenty-one or eighty-five. You have to satisfy the soul, to satisfy yourself.'

39

When words arrive out of one's soul, then one feels a certain release, and whoever offers that release, towards him or her one tends to become attracted.

Every individual has secret weaknesses, and the poet had a weakness for those who displayed intelligence.

(Intelligence is not merely cleverness. It is a blend of thoughts and feelings, intellect and heart.)

The poet felt an eagerness to meet the virgin; the lure of their conversations attracted him. Even though a depressed mount of Venus on the virgin's hand, suggesting a weakness in matters of pleasure, had disappointed him, she impressed him like no other.

The virgin possessed a reserved nature. Even those who were supposed to know her, didn't seem to know her at all. The group of students with whom she hung around, had little idea of her deeper thoughts. For the last two years, she had bored herself with fun and entertainment, and the boredom had surfaced as a look of indifference and laziness.

And then, the poet appeared in her life, and brought with him, the intensity that comes with seriousness.

Very quickly, both of them discovered the sense of release, trust and understanding, that tends to develop when minds are compatible.

They took to each other's company; their friendship matured within days.

40

When the virgin had first seen the poet she had mistaken him as a skirt-chaser, a smooth talker, a smart alec and generally one of those who are surrounded by girls of questionable intelligence.

But gradually she had begun to notice the articles which the poet wrote, read some of his poems in the literary magazine and appreciated the tireless work he did for the student body.

And when she read the philosophical passage in the literary magazine, she couldn't keep herself away from speaking to him.

She was also quick to spot that the poet knew that the best way to impress an intelligent woman is to make no effort to impress her.

The poet stayed natural, honest and straightforward, and she liked that.

41

The virgin was attractive; her boredom and laziness also made her appear sensuous. Many boys and young men had tried to make a move, but she had rejected all of them and waited for 'someone better'. (She also secretly feared that 'someone better' may elude her forever, and like it always happens, she might be forced to compromise with 'someone not that better'.) Once she had felt a certain chemistry with a fellow passenger during a flight, and she had forever regretted not being brave enough to ask him his number. With that memory she sought a similar experience.

Despite her sensual appearance, the attitude of the virgin in matters of love was marked with prudence and caution. Confessions of some of her acquaintances, who seduced sixteen-year-old boys and allowed themselves to be seduced by forty-year-old wine bar men, shocked her.

When she heard from a secret source that the poet had helped one of her hated acquaintances to find a reliable doctor for an abortion, she had asked, 'Why did you get involved in such a dirty thing?'

'Helping out a person who is in trouble isn't a dirty thing,' the poet had replied. 'The boyfriend of the girl had come to me for help. He was very confused and afraid. He was thinking of marriage. But the girl was less sentimental, she wanted an abortion. So I helped them find a reliable doctor.'

The virgin saw a fair point in the poet's explanation. But she couldn't accept such reckless sexual forays, even if it was an act of love, and not of pleasure alone.

One day at a bakery near a spiritual commune, where the poet and the virgin went for Danish pastry and watermelon juice, she suddenly asked the poet about the number of girls he had done 'things' with. The poet smiled and then told her a single digit number. 'What's your target?' she asked. The poet paused for a moment and said, 'To develop myself, to write poetry and to find answers to old questions.'

The poet's reply embarrassed the virgin. She was about to fall into silence, but the poet changed her mood.

'Since you asked me a very personal question, I have a right to ask you one,' the poet reasoned, 'You asked me how many girls did I sleep with. Now tell me, how many guys have you kissed?'

The poet burst out laughing. The virgin felt like slapping him. But she had never slapped anyone, nor had she ever kissed a guy, leave alone make love. 'You already know, don't you? Why are you asking me then?'

'Perhaps, I shouldn't have...' the poet apologized.

A young foreigner, dressed in a maroon robe, politely asked for the poet's lighter. The poet lit the cigarette for her. Then he offered the virgin a cigarette. She took one, shred it into pieces and met the surprised look of the poet with a coy calmness.

The poet didn't like the utter waste of a cigarette. He tried to explain that it was not fair to 'kill' a cigarette before it could fulfil its destiny.

'Nice words, but I rather like it when you become angry,' said the virgin.

The poet smiled. 'But I don't like being angry,' he said. 'It makes one do stupid things.'

The poet dropped the virgin at the house where she lived as a paying guest — sharing her room with another girl who had terrible toilet manners. *Don't react so strongly to trivial things*, she could almost hear the voice of the poet, when she found out that her irritating room-mate had left without caring to flush.

She also recalled what the poet had told her: *In life, most people will disappoint you, but that shouldn't matter. You only have to make sure that you don't disappoint yourself.*

The poet's words always made her think and made her aware of a deeper part of herself, even though she often disagreed with him only to spark a debate.

But she had begun to deeply value his company and sensed an eagerness within herself to spend time with him.

42

After a month, while dressing up in front of the mirror, her irritating room-mate passed a remark. 'You never took so much time to dress up before. Something is cooking in your life or what?' She told the room-mate that nothing was 'cooking' in her life, she just felt like wearing the clothes she had never cared to wear before.

At that moment, the virgin became aware that she always let her hair loose and took care to look her best before going out with the poet. Something started to bother her.

For the next few days, whenever she thought about the poet or went out with him, the other person who came to her mind was the poet's girlfriend. She liked her, even though the sweetheart didn't like her and never spoke to her nicely. She understood that the sweetheart perceived her as a competitor, an enemy, and hence behaved coldly with her. Then the virgin realized it wasn't fair of her to occupy so much of the poet's time. An undercurrent of tension had also developed between them, as if they had begun to expect something else from each other. She liked the anticipation – the sense that something might happen – but it also made her nervous.

Nervousness is caused by expectancy, and when expectancy disappears, so does nervousness.

The nervousness of the virgin vanished when she decided to hold herself back and stop spending so much time with the poet.

Now, she felt a bit sad. Fearing the insights which might surface, she didn't want to probe into her sadness. Once she had decided on something, she knew she had to stick to it. She had to avoid being swayed by selfish emotions, and be brave enough to do what was fair, right and just.

43

Before the poet could feel that the virgin had begun to hold herself back, an insight occurred to him during a stag party.

Around fifty young men had congregated in a flat to celebrate the birthday of a boxing champion. Vulgar jokes, wild dancing, loud music, strong smell of joints, eye-stinging smoke, well-cooked meat and alcohol dominated the atmosphere. Everyone targeted a skinny lad who had a reputation of getting drunk after merely inhaling the smell of whisky. Nobody was surprised when he got drunk after drinking half a bottle of beer and started leaking out his secrets. He said something that the joker modified to suit the mood of laughter. 'Hear what he just said,' the joker cried out, 'a whore charged him double for taking too long to climax.' Everyone burst into a frenzy. More beer was poured out.

A little while later the skinny lad began to talk of his love life. He confused the name of his girlfriend with the name of another girl who infatuated him. If his girlfriend's name was Sita, and the other girl's was Emma Bovary, the skinny lad declared, 'I love Sita Bovary.' Everyone erupted once again.

But the roar of laughter couldn't distract the poet from spotting the complexity that had been flushed out by alcohol. It was possible that some of the traits which he loved, and felt attracted to, were found in one girl, while the other girl possessed the rest. Hence, the skinny lad had allowed his wishful imagination to create a third character by mixing the traits of the two girls, who appealed to his head and heart.

When the thought occurred to the poet, his mood changed. He lost his interest to remain at the party. He started to think about the sweetheart and the virgin. And like always, he preferred to think in solitude.

Away from the noisy party, the poet didn't notice the relief of the quiet street. He walked by the midnight buildings – silent and dark – and the street dogs who allowed him to pass in silence.

Thoughts of the sweetheart offered him a weight – a tight sensation around his throat – that lessened, when the virgin arrived, to occupy his mind.

44

Different people stir different aspects of a person, which prompt him or her to respond in different ways. Someone may reach those aspects of the self that remain largely untouched by any other.

In the context of the poet, the virgin had been able to gain access to the concerns of his intellect where the sweetheart didn't penetrate – she believed more in emotional aspects – intellectual concerns didn't interest her.

But sometimes, love for a common ideal succeeds to bind people together more than the love they might have for each other.

The sweetheart was close to the poet's heart – she inspired love poems; but the virgin had come close to the idealism of his poetry – his concerns about life and the times.

The sweetheart stirred the poet's heart, while the virgin stirred his mind. Both touched him, but in different ways.

45

Once the poet reached his room, he realized that in the past couple of months he had drifted towards the virgin, without ever realizing how much he had neglected the sweetheart. Reproach filled his mind.

He was being unfair to the sweetheart, and wanted to redeem his negligence.

The sweetheart monopolized his mind. He felt a sudden surge of love for her. He wished to caress her, hold her and kiss her. He wanted to love her like he had never loved her before.

He was overcome by a torrent of longing. It confirmed to him that his heart lay with the sweetheart, and the virgin was a distraction, a friend.

46

Abhiman is an Indian word of which no exact translation exists in the English language. But an attempt can be made to explain the 'state of experience'.

One can say that soul is egoless, while ego is soulless.

In other words, the experience of ego is the absence of soul-sense, while the experience of soul is the absence of ego-sense.

Abhiman is a state of experience where the soul is overcome by the ego.

Abhiman arises between two lovers usually due to a lack of understanding and those things which are perceived in hindsight as silly trivialities. (All serious conflicts of love ultimately appear childish.)

For example, on one occasion when the poet learnt from the joker that the sweetheart had made plans for a party and the poet was expected to join the party on a particular date and time, he was immediately engulfed with abhiman. The poet expected to be informed of such a gathering beforehand. He expected that his sweetheart would inform him before she informed the others. He didn't appreciate that a rank outsider, even if the outsider happened to be a close friend of his, had informed him about the plans of his lover. The poet felt ridiculed in the eyes of others (to not know, what he should have known first) and let his ego wreak havoc on his soul. He suddenly perceived an alienation from his sweetheart that caught him completely unawares.

What did the poet do in his state of abhiman? He allowed the will of the ego to hold sway over the will of the soul. He avoided meeting the sweetheart, switched off his phone, refused to grace the party hosted by her, and distanced himself from his friends who thought that the poet's behaviour wasn't justified.

But abhiman manifests itself as rock-like on the exterior, but beneath the surface, it is delicate. Despite the frigid pretence, the lovers hope for a gesture that would enable them to break away from the terrible state of abhiman and free themselves from their suffering.

In the poet's case, the sweetheart had taken the wise step to end the cold war of hardened egos. She had managed to clear the misunderstanding. They met at their favourite restaurant, the communication between them was finally restored. The spell of abhiman broke. The nourishing waters of love, emanating from the soul, broke the dam constructed by the ego and flowed triumphantly without any hindrance.

47

'Did he give you a call?' the married woman asked the sweetheart in the room where the instructors changed their clothes and exchanged opinions.

The poet never shared any detail of his love life with any of his friends, but the sweetheart had a trusted coterie of friends in the fitness centre and they took great pleasure in discussing every turn of their love lives.

In the small changing room, where mirrors on the walls added a false impression of space, the sweetheart had a lot to talk about and never lied about the facts. She also felt proud when the other girls mentioned good things about the poet and held him in high esteem.

But as her love deepened with the passing of time, she discovered that her friends were only concerned with the shallow aspects and there was no one worthy enough with whom she could talk about real love, except the married woman who was six years elder to her.

The sweetheart didn't attend her college lectures. She was kept busy by her extra courses at Max Mueller Institute which she had to pass in order to get the scholarship that she aimed to acquire. But she never missed her dates at the fitness centre and felt free to confide in the married woman whom she had begun to think of as a friend.

The sweetheart despaired to the married woman that the poet had simply forgotten her; she couldn't find the poet in the campus and the poet never thought of giving her a call. (The poet had the habit of keeping his mobile phone in the silent mode, and more often than not kept it switched off to avoid being irritated by needless calls of gossip and SMS jokes.) And for the last one month, feeling the affliction of abhiman, she had taken a vow to not meet the poet, or call him, but had waited anxiously for his phone call.

The married woman had revealed a caring nature. But when she learnt that the poet still hadn't called, she changed her stance. She took upon herself the role of a 'well-wisher' and offered many ideas which questioned the sincerity and the fidelity of the poet, and which in turn afflicted the mind of the sweetheart with needless doubts and despair.

'He is taking you for granted. If you let men take you for granted, you are doomed,' the married woman said.

'What should I do?' the sweetheart asked. She expected that her friend would tell her to remain patient, advise her to forget about her silly vow and ask her to find the poet, speak nicely to him, as if nothing had happened.

'Forget him,' the married woman said. 'He is not the sole man left in the territory. I will introduce you to a cousin I have. He has a great future. He is into computers. He will surely settle abroad. I am sure you will like him. He is just your type.'

The sweetheart looked at her in utter disbelief. The suggestion was blasphemy to her; she couldn't say a word. But her face said it all, 'You silly idiot. It's not about finding a man to pay my bills. It's about the man I love.'

A girl came into the room, greeted them, picked up a bag and left. The thumping of feet and stereo music came in loudly, till the door slowly closed by itself and muffled the rhythmic noise from the hall.

'I am only thinking about you,' the married woman continued, 'probably he doesn't love you anymore and...'

'No. He loves me,' the sweetheart said and paused for a moment. 'Even if he doesn't love me, I love him, and will continue to do so, no matter what.'

'Great. He will spent most of the time with that bitch. God knows what they may be up to!' the married woman blasted. 'Occasionally, he will remember you when he wants to sleep with you, and my good sweet girl will be happy and pleased.' The married woman's tone was that of hopeless ridicule.

'Don't speak like that,' the sweetheart shouted, and regretted that she had ever told her about the virgin. 'I made a mistake by telling you about his friend,' the sweetheart said as she quickly stuffed her bag. 'You don't know him. He might cause me pain but he will never cheat on me.' Then she hit back, 'Start worrying about the secretary that your husband might be sleeping with.'

She bid a loud unfriendly 'bye' and rushed out.

48

The sweetheart changed her dates and timings at the fitness centre – she didn't want to be accosted by the married woman, see her crafty face and listen to her repulsive suggestions. Now it struck the sweetheart that she knew so little about the married woman, but she had disclosed the gravest despairs of her life to her. She blamed herself for being so naïve. Then she realized why she had trusted her. She had been feeling the weight of all the unsaid things which had accumulated within her, and she had longed to relieve herself by confiding in someone who would listen. A paltry bit of caring attention and silent nods of agreement had offered her a blanket of comfort and assurance.

But she had made an error of judgement – she had been too quick to trust the married woman. She felt unhappy, and remembered the touching story that the poet had told her: the poet knew a woman who was the mother of a friend he had in high school. He never went to his friend's house fearing that his mother – a homely woman in her mid-forties – would begin to talk and simply wouldn't stop. Sometimes when she called on the phone to inquire about trivial things, the poet never managed to hang up before an hour. The woman did not seem to realize that boys in high school would have no interest to know about the problems she had with her maid, the songs which she liked twenty years back, and what kind of spices should be used to cook mutton in summer. It was a torturous experience for the poet.

But he had also realized a deeper truth. The woman had no one to talk to – her husband and her son stayed within their own spheres – and once she found someone, she simply went on talking, and didn't want to stop.

The sweetheart now thought, deep down her story was also the same. She had wanted to talk from her heart, but there was no one to talk to. There was no one who would listen. It was a sad suffering, she thought, that every person has to face.

49

The sweetheart returned home from the Institute, retreated to her room and changed. It was six in the evening when she finished taking her hot bath. She thought of visiting the café where the poet usually hung out in the evenings. She decided against it feeling once again the pangs of abhiman. But whenever the will of the soul is overcome by the will of the ego it produces unhappiness and discontent. The sweetheart's mood worsened. Even her dog failed to make her happy. Something started to boil within her. She succumbed to an outburst. She shouted at the maid for getting her a mug of 'horse urine' instead of coffee.

But the maid argued that the coffee was exactly the way the sweetheart usually liked it to be. A shot of anger rose within the sweetheart; she restrained her impulse to break the mug but slammed a door. The noise hurt her eyes, more than her ears. She winced painfully, and immediately felt a wave of regret. She called the maid, spoke nicely to her and drank the mug of warmish coffee.

The sweetheart inquired whether the maid's husband – jobless and an alcoholic – had come to demand money from her, whether her old parents were keeping well in the village and other such things.

Before the maid left the room to take care of dinner, she said, 'It's good in a way that you at least remember me, after you scream at me.'

The words of the maid suddenly made her think of what the married woman had told her: 'He will remember you when he wants to sleep with you.'

The remark of ridicule filled the sweetheart with unhappiness. Looking out of the window she began to doubt the poet's actual need for her. Probably what the married woman had said was right. A strange despair filled her. It pained her to think that the poet might have no other need for her except sex.

With the thought something relented within her. The word 'sex' had never before occurred in her mind. It aroused the same disgusting feeling that she sensed when she had to look at her used tampons – it made her stomach flutter and filled her with sickness. Terrible thoughts pranced around her mind. When she thought herself as the poet's partner for pleasure she felt betrayed. A shot of pain and despair rose within her. Then she realized how much the word 'sex' threatens love, devalues a relationship and puts love under suspicion and threat.

Then she thought, if the poet in her friend's apartment switched on the television set without making an advance towards her, she would feel humiliated and unwanted.

Then why did the word 'sex' stir those feelings within her, the sweetheart questioned herself, and retreated from the window.

She drank a glass of water, put on some slow music and snuggled into bed with her dog.

A little while later things became clear to her. She perceived a vital difference between 'sex' and 'lovemaking' – sex is the appetite of the body while lovemaking is that of the soul. Sex is the crude desire to derive, while lovemaking is the tender desire to give. She had never had sex with the poet, she had only made love.

She now understood that the remark – 'he will remember you when he wants to sleep with you' – meant the poet didn't love her. And that had caused her distress.

The sweetheart, once again, felt weighed down by the fact that the poet didn't offer her enough signs and gestures to announce his love, and his actual need for her.

The sweetheart kept on brooding – the chance remark of the married woman had succeeded to gash out a fresh harvest of doubt and grievance.

Next afternoon when the poet, overwhelmed by his love for the sweetheart, went to meet her during her break at the Institute, and apologized for his negligence, the sweetheart remarked, 'So, now you care to remember me? You must be thinking of sex, isn't it ?'

50

The accusation hit the poet like a slap. The sweetheart realized the importance of her words almost immediately – she knew she had hurt the poet by the look on his face. 'I don't know why I said something like that,' she said in a flurry of distress. 'That damned woman ... I didn't mean it, please believe me!'

After a tedious minute of overlapping words, the poet was able to make himself heard. 'It won't be a good idea to talk at this moment. Things will get worse. Let us not see each other for a while. I will talk to you later.'

The poet got up from the table and walked out of the cafeteria.

Seconds later, he came back to the table.

The sweetheart looked at him with hope, distress and apprehension.

Without looking at her, he kept a fifty rupee note on the table and used an empty glass as a paperweight.

'For my coffee,' he said with a weight in his voice and walked away.

51

The neighbourhood of the Institute carried the look of lifestyle magazines. The poet walked past the well-designed buildings, guarded gates, parked cars with chauffeurs, clean pavements without cracks, and came to a crossing.

He stopped outside a fashionable pub whose exterior looked bland, since it was afternoon. The poet walked in and seated himself on a bar stool, ordered a pint of beer and lit a cigarette.

The poet was keenly aware of the hurt he was feeling – he had been accused of stooping low to serve his interest. He didn't notice the familiar music that was playing softly, neither did he survey the mirrors in front of him, to inspect the other customers who occupied the tables behind him.

The young bartender, probably a trainee, wasn't used to seeing reflective faces with faraway eyes, clouded behind the hazy smoke of cigarettes. He only knew the commerce of fun – he assumed anyone who didn't smile, or sway a bit to the music or looked around, was unhappy.

'I can make a cocktail that would cheer you up,' he said, with a swagger of forced exuberance.

The poet was distracted. He looked up to meet the smiling face of the young bartender.

After a pause, the young bartender repeated his remedial proposal. But the poet didn't want to be cheerful.

Without saying a word he gulped down the beer, kept a couple of notes under the crystal ashtray, picked up his pack of cigarettes, and walked out.

52

Back within the habit of his room, the poet could finally release himself.

The blunt remark – you only think of me when you think of sex – suggested, he only sought to use the sweetheart.

It meant, he was exploiting her, under the pretence of love.

Once again he felt wounded. He had never viewed the sweetheart as a partner for pleasure alone. In fact, he could only view her with his heart, with feelings, with sentiments.

But now, his mind was tired of his heart. It pointed out what his heart had tried to suppress. The sweetheart wasn't mentally compatible with him. And that was the truth – she bored him, and he could no longer rely on spurts of sentiments to sustain a relationship between two individuals who weren't intellectually compatible.

It's tough to face truths. The poet felt uneasy. His heart weakened. Then he told himself, he was overreacting. His mind was disturbed. He should wait, and wear off the anger and the hurt.

53

With the progress of the evening, the shadow of the table lamp, situated beside the south-facing window, gradually lengthened – the slanting shadow continued on its journey eastward, while the sun was already diffusing behind the hillocks in the west.

The poet was at his desk, staring at the lines he had just written on his laptop:

> Far too long, we were in love,
> not knowing each other.
> Far too long, together we have grown,
> far away from each other.
> Far too long.

The poet stared at the words. Poetry never lies. They had drifted apart, and didn't understand each other anymore. They made each other suffer. And that was the sad truth.

54

The passing hours had smothered the ill-feelings triggered by the sweetheart's remark. He also read the series of SMSs from the sweetheart. He understood that it was an impulsive outburst and the sweetheart had said something that she really hadn't meant. He too had neglected her, and that had forced her to harbour doubts about his sincerity. Everything had combined to produce yet another misunderstanding.

All of a sudden, he felt tired of both – to explain and to seek explanation.

Demons of misapprehension always tend to come back with a renewed fervour, like those viruses which become immune to vaccines, and return with more strength and complexity. The pattern remains at work forever – there is no permanent solution, only a permanent battle.

The possible outcome of such a recurring mishap was known to him – he had seen it from close quarters in the case of his parents, the damage it can cause within individuals, and then, within their relationships.

To love someone, and to be in a relationship with that someone is not the same – the former is much easier than the latter.

The weariness of every relationship – the continuous battle, the tedium of overreactions and the increased occurrence of misunderstandings, made him sick.

He had had enough. He was not going to mark his time on this earth by battling to clear misunderstandings!

55

Emotional decisions or the decisions taken under the influence of full-blown emotions are always short-lived and never journey the distance. A mind afflicted with anger or sentiments may decide hastily, but once the waves of emotions recede, it becomes difficult to execute what was decided upon.

The poet had seen this phenomenon occur repeatedly with his parents. They would get carried away by emotions and declare their decisions vociferously on the spur of the moment. But after a few days, when their minds regained composure, they themselves wouldn't see any sense in what they had declared to themselves and to others.

Reverse learning was at work once again. Everyone around him displayed the examples of what not to do, how not to behave.

From the days of his adolescence, the poet tried to cultivate the habit of taking his decisions when his mind was balanced. He had noticed, wise decisions are only taken when the mind is calm, even and well-motivated; unwise decisions are always made when the mind is disturbed and agitated. His readings of spiritual philosophies had begun to serve him well.

He realized wise decisions are also tough decisions. One has to rely on grave feelings which don't clutter the mind, don't weaken the heart. They are sublime feelings of intelligence, of wisdom and of mental strength.

He came to the conclusion that a relationship of love is governed by four types of compatibilities – physical, emotional, intellectual and spiritual. These four compatibilities, apart from circumstances, determine the success of any relationship.

When he thought about his relationship with the sweetheart, he understood that the body and heart are not enough. Such a relationship can only be sustained by compassion and sentiments. In time, it would weaken, sufferings would begin to unfold, it would become a burden, a mistake and a regret.

Now he felt, he had to save himself and the sweetheart, from each other.

He had to save them, from the sad situation that had come to exist between them – compatibility of body and of heart, yet, incompatibility of minds.

He was not prepared to go ahead with this jumble of evenness and roughness, equality and inequality, union and divergence.

With a heavy heart, but a still mind, the poet understood, what decision he had to take.

56

After two months, when the university examinations finally got over, the poet spoke to the sweetheart in her friend's apartment.

'Are you in love with someone else?' asked the sweetheart. 'No,' replied the poet. 'I just need to be freed of the commitment that has grown between us. I want to be left alone. I want to perceive myself from a distance.' His words were not rehearsed.

Tears started to fall from the sweetheart's eyes. Sentiments rose in the poet's heart. His heart implored him to embrace the sweetheart, take back his words and make fresh promises. But his mind didn't betray him; it remained firm against the torrent of his sentiments.

The strength of a man is often tested by the tears of a woman. The poet, on that occasion, managed to hold on.

The sweetheart controlled herself with dignity. The decision of the poet was truly not shocking to her. Somehow she had prepared herself for the day when the poet would finally decide to leave.

But the sweetheart didn't feel dishonoured – the poet was not leaving her for any other woman. She ignored the words of her friends who had told her, a man doesn't leave a woman unless he has found another one. She knew that the poet cannot lie. He said he wanted to be alone, and she believed him.

The sweetheart had acquired the scholarship. She was all set to leave for Germany in four months. Secretly she thought, when she returned after two years, the poet might have changed his mind. And that cushion of hope mollified her distress.

Like always, the sweetheart made coffee. They drank in silence. Then the sweetheart suddenly grabbed the poet, looked straight into his eyes and told him. 'You must do two things for me. First, you will have to give me your T-shirt. This one that you are wearing now. I have bought a new T-shirt for you. I will give it to you. And second, you are going away tomorrow, and perhaps forever. Make love to me for the last time.'

57

The sweetheart's words alarmed the poet. He was in no frame of mind to fulfil her second wish. 'I cannot do that,' he said. 'I am sorry.'

'Why not?' the sweetheart asked, and came close to the poet. But he moved her away.

Then he picked up the new T-shirt from the corner table, changed, and gave his grey cotton T-shirt to the sweetheart. 'I think you should wash it,' the poet said. 'It's smelling of sweat.'

The sweetheart smiled. 'Don't you understand? I will never wash it,' she replied. Then her eyes became moist and vacant.

58

Wearing the new moss green T-shirt, the poet decided to take the longer walk from the apartment to the campus.

The melancholy of the farewell slowly surged within him. By the time he arrived at the familiar tea stall, near the west gate of the campus, he was choked with emotions.

At that moment, an acquaintance came to accost him. A conversation followed, tea and cigarettes consumed.

The acquaintance left after twenty minutes; the poet discovered that the melancholy had subsided.

He looked at his watch. It was nearing eight. He had to keep a date. It was time to bid the virgin his farewell.

59

'What are you planning to do next?' the virgin asked the poet at the bakery near the spiritual commune.

'I don't know,' the poet replied. 'Probably apply to a newspaper.'

The virgin was headed to a business school. She had felt surprised that the poet didn't want to do the same.

The poet explained to her his circumstances. He already had enough well-being – an apartment, a car and term deposits to take care of the monthly bills.

His mother was dead – there was no burden of expectations; he never spoke to his father, who lived in a different city, and his brother, who had gone to America. His maternal grandparents, who also lived in another city, didn't interfere in his decisions.

So it was matter of doing what he loved. He loved the written word. And that was it.

Then the virgin asked the poet about the sweetheart. The poet told her the sweetheart was headed to Germany to become a foreign language teacher.

'So you are going to wait for her?' the virgin asked.

'No,' the poet replied. 'We have broken the commitment.'

'Why? What happened?'

'I want to be alone. Don't want to rush into living a life whose worth I haven't realized.'

'Or is it that you have someone else in your mind?'

The poet smiled. 'No. I don't. I am tired of relationships. I need to organize myself. Catch up on my reading, my poetry, my inquiries.'

The virgin smiled at him. 'You are so different. I hope you will be happy. I hope you will keep in touch.'

'But you know,' the poet said, 'we might keep in touch for a while. After that, we will have things to do, new people will come into our lives, emails and phone calls will dry up, and we will even forget to follow each other's Facebook updates.'

The virgin looked sad. The poet was right. When and how they would meet again, they themselves didn't know.

After the meal, they took an auto. The music kick-started with the engine. The Chinese audio player and the locally made speakers – with hissing treble and high bass – made a dreadful noise. The poet had to shout to the driver to turn it off. The music stopped, but the speed increased. Within twenty minutes, the vehicle came to a slow halt at the gate of the house where the virgin lived as a paying guest.

Finally, it was time, the moment of parting had arrived. 'How I hate these goodbyes,' the virgin remarked.

'You better get used to them,' the poet said. 'Every hello has a goodbye at the end.'

The virgin smiled. The poet gave her a hug. She stood at the gate, and watched him leave.

60

The poet returned to his hostel room. Tomorrow he was leaving for home. He had opted out of the graduation ceremony. He had marked the other option; he would receive his graduation certificate, marksheet and the degree by registered post. There was no need to return to this city anymore.

The poet looked around the room – it had been his home for three years. Now it looked bare – everything had been packed, except the towel, the toilet kit, and a few books and notebooks which lay on the table. The bloated backpack and the two suitcases waited in the corner. All the posters and the art work still remained on the walls. The poet didn't bother to remove them. He left them for the next occupant.

The poet's friends were waiting for him. Everyone was nostalgic – a memorable period of their lives had come to an end. And now they had to leave the safe refuge of the hostel, go out in the world and survive.

The joker arrived with his usual cheer and offered a joint to the poet. 'For the sake of the joint family that is about to disintegrate.'

The poet smiled. He took a puff. But grass had lost its appeal and grown stale – he no longer wanted it. Now I only need a clear mind, he thought, and handed back the joint.

The poet's train was scheduled to leave the next evening. There was enough time for the poet and the joker to reminisce. They walked out to the passage and leaned over the railings. The courtyard appeared desolate. The lamp post stood in darkness. There were clouds in the sky, the breeze smelt of rain.

'I might take up my uncle's proposal,' the joker informed. His uncle, a bachelor, lived in a popular Himalayan hill station. He owned a couple of mid range hotels and a travel company that specialized in adventure tourism. He wanted help and the joker was his choice.

'That would be a good life,' the joker said, 'living in the hills, with an endless supply of grass.'

After a while, the joker said, 'I somehow cannot forget the moment when you pointed to a bird and said look time is flying. It has stuck with me. Whenever I see a bird, I see time fly.'

But there were no birds to be seen at that moment. The crows also seemed to have vanished.

'True,' the poet agreed, 'time has flown away.'

61

Next morning, when the poet was packing all the remaining things into his backpack, he paused to flip through the various notebooks containing poetry, quotes and passages from various books, and the insights which had flashed in his mind.

After a while, he packed all his notebooks except for one titled 'Seasons of Time'.

Four years ago, the poet had discovered the phrase and wanted to understand its meaning. So he had bought a hardbound notebook and had written that phrase on the first page. His strategy was to fill the pages with ideas, quotations and insights, which might help him to understand the phrase.

He carried the notebook to the college canteen and ordered breakfast and a cup of tea.

While going through the notebook, the poet realized that he had continued to fill the pages with a marked regularity.

A few months ago, he had made the following entry:

'Every lifespan, from birth to death, is marked by distinct seasons of time – phases and periods – which an individual encounters and passes through in life.

'But what are the ingredients of such seasons?

'Events, things, trends, incidents and situations.

In other words, the combined experience of people, places, things and circumstances.'

Those are the ingredients of time, the poet thought, people, places, things and circumstances, and the manner in which one responds to them.

62

In *Kaushitaki Upanishad* (I, VI) there is this line: *I am a season, son of seasons.*

The poet had written down the line in the notebook that he was now reading. He came across the quote and paused over it.

The self has been spoken of as a season – as an *experience* that has been formed out of *experiences*.

And this *experience* continuously evolves through varied *experiences*, and changes itself.

In other words, the *self evolves through itself*.

This is how it happens – triggered by the seasons of time, the self evolves through feelings, thoughts, speech and actions.

The insight satisfied him. The evolution of self is an innate design of life.

Then he thought, that in life, in reality, one cannot experience anything but one's own self.

All the experiences of people, places, things and circumstances only happen in relation to our own self, and depend upon our own response.

It's very true, the poet thought, that the beauty of a painting is dependent on the self that experiences it. The reality of pain is dependent upon the self that has to deal with it.

Life is a stage for self-experience and a stream of feelings, thoughts and actions which flow from that experience.

The poet felt joy when the thoughts occurred, and he quickly wrote them down in his notebook.

After a while, as he sipped on his fourth cup of tea and smoked a couple of cigarettes, he thought:

Now the question is, what exactly is the self? What is its framework, its design?

The poet opened the notebook and on the last page he wrote:

The human self has come before all religions, nations and boundaries. What is the self? This is the question.

63

The train jolted into a slow start within the cacophony of the railway station.

The poet had finally managed to put his luggage under the berth after a tiring argument with a fellow passenger who was travelling with four enormous suitcases.

But he had enough time to rush back to the door.

The joker saw the poet standing near the door of a compartment; he was waving at him. He waved back.

The train receded to a distance. He remembered two lines which the poet had told him twenty minutes back.

> Life is a river formed by two tributaries of
> beginnings and ends.

64

The poet was returning home to both – a beginning – his life after college – and an end – the finality of one summer morning, two years ago, when his mother didn't wake up from sleep.

It happened during the summer vacation, after his first year of college. The poet was visiting his mother. But she wasn't the mother he had left behind.

Under the softness of the reading lamp, she would be lost in her thoughts. Her short life lengthened by bitterness would converge in a moment. Her face would appear cold and frozen, her eyelids would refuse to blink; she would be trapped within memories, which haunted her, caused grief and drained her of life.

His stepfather, whom he had never met, had passed away suddenly. His mother had gone into a shock. She didn't try to recover. She had lost the will to live. She suffered a cardiac arrest and died in her sleep.

The poet was not yet twenty at that time – he had never been affected by death – it was a curious experience. He felt severed, vacuous, his knees went weak, his hands trembled when he had to call his grandfather.

His mother was his family – the link of womb, a string of self. When it snapped, it revealed a strange emptiness, into which he drifted slowly, like a weightless feather, with the enormous weight of his grief.

Everything became hazy after that. People came to visit, the apartment became crowded, ceremonies were organized. The poet had to light the final fire, chant the words, immerse her ashes in water.

Time passed like a dream. His mother became a memory, a remembrance – a few photographs in which she smiled.

65

The train had settled into a monotony of speed, and the poet sat by the window. He had his MP3 player on, but he wasn't listening to the music. He was coming to terms with another end. College life was over. Everything and everyone, who had been close to him for three years, had vanished and disappeared. The sudden void was filled with stirring sentiments. The night beyond the train window, the breeze against his face and the rush of memories almost choked him.

He rushed to the jerking toilet, balanced himself and splashed cold water on his face. The touch of water on his skin managed to slow down his pulse. He cleaned his face with liquid soap and found the courage to face the mirror. He looked grave, tired and unhappy.

He smoked a cigarette in the toilet and went back to his seat. A fellow passenger tried to speak to him and offered him a film magazine. 'I don't read them,' the poet told him, and plugged in his MP3 player.

He stared out of the window. The moving train and the music soothed him. He found the space in his mind to think about something else.

Time came to his mind, and that question he could never answer – what is time?

But now, words arrived in a string to make a sentence, with a rhythm and a wave: *time is experience, experience is time.*

He thought over the sentence – it satisfied him. Time and experience are inseparable. There is no time without experience, and there is no experience without time.

Then he thought: Time, self and experience are woven together. Seasons of time, also mean, seasons of the self.

Time is not only determined by what happens to us, but also by what we make of it, he thought. This is a period of transition. A season has come to an end. A new season will now begin.

Two

Two

Myriad Void

0

The girl was wearing lipstick of a shade of orange. She was lying naked and in position – waiting for the manoeuvre of penetration – on a moist hard bed. At that precise moment, the air, condensed by the smell of dampness, heat and desire, was choked with the loud noise of a religious prayer, amplified by an electronic sound device.

The announcement of the prayer alarmed the girl. She grappled for the sheet of cloth that had fallen on the floor by the side of the bed, fiercely shook her head, pushed the poet away, grabbed the sheet and covered herself with the loud print of flowery motifs.

Then she indicated towards the direction of the prayer with her eyebrows and head. The poet understood, she was unwilling to have sex while the prayer continued from the crest of a nearby minaret. The poet pulled away to one side of the bed. The girl made a gesture with her hand – it suggested, 'we have to wait'. The poet gave his nod of consent.

The voice filtered through the closed wooden windows and the destitute bricks of the old building. The room was large and had a high ceiling. But it was partitioned into several small cubicles, furnished with narrow beds and loose hooks on the wall.

Each cubicle was so small that a 40-watt bulb – without a shade – could nearly replicate the heat of the summer sun.

The warmth added to the smell. The odour of sweat, dampness and cheap perfume became more dense. The prayer added a new ingredient – it made everything very sombre, and very grave.

The poet waited by the side of the girl. She had covered herself well. She lay on her back and rested her head on a small grimy pillow. She looked above at the small ceiling fan that was whirling with a squeaky noise.

The dusty maroon floor was littered with cigarette ends. At the corner, near the sliding door, dark brown pieces of a smashed beer bottle had been piled up along with dirt and a dead cockroach that was infested with tiny red ants.

The army dissected the larger insect. But they didn't eat it. They carried back the organic fragments into their crevice to store for winter. (The cockroach, despite being trampled on mercilessly, had crawled bravely, leaving a short liquid trail, before succumbing to its injuries.) One broken piece of caterpillar-like dirt, created by the edges of filthy fan blades, had got stuck in a spider's web, swaying at a corner of the whitewashed walls.

The poet's amorous adventure was oddly suspended. He gazed at the spider's web. Transfixed by the unexpected event, he waited patiently for the prayer to get over.

1

Then a drama erupted. The poet till now had responded to the situation with his instinct. He didn't have the time to feel and grasp the situation. During his wait, his breathing had slowed down, his urge had suspended, and his mind had calmed to allow the insight to surface. In a flash he grasped the situation. The waiting poet – wearing a crumpled condom over his limp penis, the stiff religious girl – prepared to protest furiously if the poet lost his patience, the squalid cubicle of the brothel, and the loud voice of the prayer, suddenly made him realize the absurdity of the situation.

At that very moment the light flickered, and the fan jolted, due to a sharp fluctuation in the voltage. He felt a joyous outburst. A few minutes ago, breathing violently with desire, he was about to penetrate the girl when the prayer had erupted, the girl had pushed him away, covered herself, and he had agreed to wait patiently till the prayer got over, and here he was, staring at a spider's web!

How absurd was the situation! How absurd could *he be*! He couldn't stop himself from laughing at the sad situation, and at himself.

2

The girl was perplexed with his behaviour. She had no idea why the stranger beside her was suddenly laughing. She had the instinct to understand that the laughter wasn't directed at her. A few minutes ago, when the poet had chosen her from among the other girls, she could sense that the young man wasn't a threat. He was least likely to coerce her into perverse vulgarities. His presence didn't inflame fear. He was good looking and clean. He had treated her like a woman, not a whore, and in return, she had even offered her mouth and had allowed the stranger to kiss – a privilege, she never offered to her customers. He had treated her with respect, and when the prayer had begun, he had also respected her religious sentiments. But why was he laughing?

She focussed on the poet's face, and tried to figure out his expression.

He appeared to be gently laughing at a thought, smiling at a realization. His face resembled a face that had recalled a happy memory.

A moment later the prayer stopped. The girl pulled aside the sheet. But the poet didn't change his position; his urge had vanished due to the cleansing effect of laughter.

Sexual desire spreads slowly as thin currents of feelings. It accumulates bit by bit, changes the feel of the mind, and makes it grave, solemn and serious. That in turn, makes one aware enough to stand up to the desire, like a humble child, standing up before a principal.

Desire becomes the authority, the power and the influence. It can withstand playfulness, but it cannot withstand lightness.

Pure laughter, emanating from the pit of the stomach, crumbles desire, disintegrates the accumulated, cuts off the current.

The poet's mind was infused with lightness – a load had vaporized. The burst of laughter had released a fresh force. He felt lighthearted, buoyant and happy.

The poet crossed over the girl, snatched his clothes from the hooks and began to dress. The girl got up and grabbed his hand from behind. She asked what had happened; the poet told her he had to go.

She made a hesitant effort to come closer to the poet but he gently pushed her arms away.

The girl sat on the side of the bed and watched the poet. When he handed her a tip, she refused to accept. 'What for?' she asked with her expression.

The poet slipped the note under the pillow, smiled and left.

3

In the dim waiting area of the brothel, the marketing trainee had become impatient. He had finished long time back – the prayer hadn't halted his pleasure. He thought about the poet and wondered why was he taking so long.

Two new customers arrived in the room. He felt a his fear intensify – what if someone recognized him? What if he recognized someone? He took out his white handkerchief and positioned himself. He would start to wipe his forehead, if by any chance, he had to cover his face.

But there was no cause for concern. The customers were complete strangers. He relaxed a bit, wiped his face and looked at his wrist watch.

The poet found the marketing trainee seated on a wooden chair. He was stiff, uncomfortable, drenched in perspiration. The poet apologized for the delay. The marketing trainee noticed the shine on the poet's face.

Both of them climbed down the narrow staircase, came to a heavy door, guarded from inside by a burly thug with a noxious body odour. The menace in his face turned to a nice smile when the poet handed him a tip.

After pocketing the toll tax, the burly thug opened the door and wished them goodnight.

4

'She was damn good or what?' the marketing trainee asked the poet when they finally sensed the relief of the street. The poet smiled and lit a cigarette.

The marketing trainee felt deflated – he couldn't think of anything at the moment that would deflate the poet. The poet had been smart – he hadn't wasted a moment to choose the pretty girl. But he himself had taken a long time to decide and had chosen the wrong woman. She had refused foreplay and made him hurry. She had also demanded a hefty tip. His sad experience was a complete waste of time, effort and money.

In the car, the marketing trainee made another observation, 'You really look refreshed.'

'Never felt better,' the poet replied and started the engine. Then he realized, he was still wearing the untested condom. Once again the absurdity of the situation made him laugh.

The marketing trainee looked at him in utter confusion. 'Did you get to know her name?' he asked.

The poet was distracted. He settled down and turned on the CD. 'No,' he replied.

'Why?'

'I didn't care to ask.'

The parking guy arrived and asked for a 'little extra' since, as he explained, the 'business is down, these days'.

The poet gave him the little extra, pressed the clutch, shifted to first gear, looked at the rearview mirror and then the side-view mirror, touched the accelerator and released the clutch.

5

The Void of Loss and Absence

After college, when the poet had finally returned to his seventh floor apartment in the high-rise building, a coarse manifestation of a void overwhelmed him. A man who can live in solitude has to be at peace with himself. But the poet was infected with desolation and discomfort. Sleep failed to refresh him, and dreams tended to punish him with nightmares.

In one such nightmare, he saw himself wandering in a ruined temple, full of burnt books and large mirrors which turned to television screens, showing a collage of violent images, whenever he came in front of them. He could feel the anxiety of not being able to find his own reflection within the depths of the mirrors. He could feel his restless desperation, as he rushed from one mirror to the other, but whenever he came in front of a mirror, it transformed itself into a TV screen.

The nightmare was so vivid that it left a lasting impression. The poet could feel a fear whenever he had to confront a mirror. While brushing his teeth or washing his face, he would feel the sudden anxiety, that the mirror on the wall may cease to behave like a mirror.

On another occasion, he dreamt of his mother crying and howling over his dead body; a waiter from his college canteen, dressed as a doctor, carrying a tray of cola bottles, lent a touch of absurd humour to the nightmare.

Home too had begun to haunt him with the remembrances of his mother. Every corner, every household item – a chair, a painting, a crystal vase – reminded him of a loss whose memories exhausted him.

The nostalgia of his past, his friends, his college, his sweetheart – everything that he had lost so quickly – enlarged the vacuity caused by the snapping of a deep emotional link, that till now had fed his soul.

Even when the sweetheart called him up, the voice didn't offer any respite but filled him with sentiments that weakened his self.

Trapped within the void of loss and absence, he longed for contact; he called up his friends, wrote them emails, and signed off each mail with the same words: time consumes everything, not like a bird of prey, but with more ferocity, like a tiger.

6

After a week, the sense of vacuity lessened. The poet rediscovered his lost self within his habits. The noise of television was replaced by the music system, a few books had been opened, cigarettes lost their bitterness and the deep windows drew him out into the distance.

Finally, he could view everything – the city, the sky and his life – with a sense of calm.

But he had to deal with irritations caused by phone calls from his mother's friends – their painful sermons and persistent inquiries into his plans for the future. He felt tired. He didn't want to think of the past, the present or the future. He just wanted to be left alone. His mind was still clouded. And somehow those tiring phone calls triggered a period of distaste, inertia and dullness. But he didn't want to sit back and suffer. He longed to escape into the world. Something urged him to seek out thrill. A face flashed in his mind, then the body. 'That's easy,' he thought.

He wanted to renew contact with a classmate's elder sister, who had made him feel that strange aching discomfort that had lasted through the night, after he had penetrated her vagina for the first time during high-school.

The poet still had her number saved in his phone. He called, but the number was invalid. He retrieved his high-school phonebook (where he wrote down numbers as a backup) from a box, flipped through the pages and found the landline number. A male voice informed him that the girl, now a young woman, was married to a doctor and lived in England. 'What a loss,' the poet thought, but he had another idea.

The poet dialled the mobile number of a high-school friend who had played cricket with him for the school team. He too had finished college and joined an FMCG company as a marketing trainee. He wanted to gather experience and letters of recommendation, which are required to gain admission into any overseas school of business management.

The marketing trainee was evidently delighted to hear from the poet. He also recalled the evening when they had gathered enough courage to pick up an attractive streetwalker to celebrate the end of high-school. 'We even shared the same whore,' the marketing trainee said in the spirit of soldiers, who had once survived a dangerous battle.

Before the conversation ended, he assured the poet in a voice of mischief, 'Trust me my friend, I know how to cure boredom.'

7

The Void of Boredom

Baudelaire wrote of two curious words, *ennui* and *spleen*. Spleen was used to describe suicidal depression whose primary cause was the soul-deadening condition that Baudelaire emphasized as ennui. Ennui isn't boredom but a more intense form of doldrums – a condition that provokes an individual into soulless hedonism. Ennui was imagined by Baudelaire as a dainty monster who smokes a hookah. Flaubert too imagined ennui, but as a silent spider, who wove in darkness, within the heart, a vague abyss of wearisome ennui.

Aleksey Tolstoy preferred to describe boredom in more poetic prose to suggest a coldness and a numbness. He illustrated boredom as 'the sister of the autumn wind, flowing from road to road, from mound to mound ... singing doleful songs'.

Void of boredom is dullness, stagnation, lassitude and inertia.

When one is afflicted with boredom the initial response is to seek tumult, thrill and excitement. More sedate the mind becomes, more gripping becomes the need for fun. It is an affliction of the mind, not of the heart. The heart is never bored, but the mind is.

Boredom has more grip on a mind afflicted with unhappiness; it creates a tendency to seek thrill as a temporary relief. Thrill cannot cure boredom it only takes the individual into a cyclical loop, a labyrinth; boredom leads to thrill, and at the end of thrill lies boredom.

A high percentage of adultery, sales of video consoles, excessive usage of social networking websites and temptations of pornography are more often than not induced by boredom. It tempts, excites, thrills, uses up time, but doesn't quite satisfy.

Long periods of time spent within the 'loop of boredom-thrill-boredom' causes extreme mood swings, irritability, shallowness, restlessness, touchy sentimentality, grave addictions and depressive tendencies.

From the other perspective, one may understand that when the mind is deprived of the nourishment of the soul, it begins to grow weak, suffers and malfunctions.

8

The poet met Boredom in a night club that he had begun to frequent with the marketing trainee who had a small group of acquaintances, all regular patrons of the club. They knew people and had contacts.

One night, the poet asked one of them about the woman sitting alone at the bar, watching the people around her. Then one of the guys asked, 'What man, you like her?' The poet didn't reply. The guy had an aura of casualness about him. He leaned back and announced to the group, 'Through the eyes of lust, every woman looks appealing.' Everyone laughed.

Then he turned to the poet, 'You don't have a bad taste, she looks sexy. But that's the deception. I have been with her once. Nothing special. She is a divorcee. She has money. Her son studies in a boarding school. Nothing to do. Simply bored with life.'

'Or, perhaps,' another guy said, 'she is depressed, and tries to escape suicide.'

'Who knows,' he continued, 'every evening, she might be thinking of slashing her wrists.'

Then he added, 'I have understood one thing in life for sure, my friends. The more promiscuous you are, the more unhappy you will be.'

The poet wasn't listening to the musings. 'Is she game?' he asked.

The casual guy smiled. 'Yeah, she likes decent young men, but the choice is with her.'

The poet's acquaintance walked up to the bar, sat on the bar stool beside her and got talking. The poet's group had occupied a table, about twenty feet away from the bar.

After a while, Boredom turned towards the table, the poet's acquaintance leaned in to add something, perhaps the colour of the poet's shirt. Lights and shadows were passing in waves and circles. The electronic dance music was becoming louder. A few shapely figures could now be seen on the dance floor.

A minute later the poet's acquaintance walked back to the table with a coy smile on his face. He sat down beside the poet. 'You have been seized, my friend. Join her at the bar. But you know the rules of a "seizure", don't you? No names, no questions, no sentiments.'

The poet knew of 'seizure', a concept popularized within the club circuit – 'Either you love or you seize. No more wasting time on dating shit.'

The poet got up, picked up his pack of cigarettes and lighter from the table. He had already consumed four pegs of vodka. He slipped the money under the ashtray, bid goodbye to his group, walked up, sat down and asked Boredom whether she would like to have another drink.

Boredom took a close look at the poet. Then she said, 'Rum and cola.'

The club was relatively empty, the bartender had less things to juggle with. He attended to the poet's signal. The poet ordered two drinks – rum didn't work well with him – he ordered vodka.

9

There is a difference between the void of boredom and the void of loneliness.

The void of loneliness is characterized by the absence of a relation, a companion, a friend or a lover, where the soul seeks a company, an intimacy, an interaction. A lonely person seeks soul-sense while a bored person seeks something to do.

Loneliness is a function of the heart while boredom is a function of the mind.

A one-night stand isn't a cure for loneliness, but a relief from boredom. It can provide the temporal thrill of excitement and pleasure, but it cannot cause soulful satisfaction and happiness.

10

The drinks, along with a plate of wafers and cashew nuts, arrived exactly when a group of noisy kids entered the club. The poet spotted a grimace on Boredom's face – she didn't like the menace of revelling teenagers.

The poet tried to strike up a conversation. She spoke in short sentences and drank too fast. The poet noticed Boredom was dull, probably unhappy. She didn't show any kind of emotion, not even a forced smile. Her eyes looked vacant, but the poet failed to notice them; Boredom avoided eye contact.

The poet wasn't keen to know, what lay within the flesh, beneath the skin. He didn't care about her mind, her nature. He was slightly drunk, his body and mind were flooded with the thrill of adventure, an urge to obtain pleasure. Five minutes later, already knowing in silence what they had in mind, she asked the poet, whether they could get out of the club. The poet agreed without a thought.

11

Boredom was a poor driver. She didn't press the clutch to its depth while changing gears – an ugly sound was generated from the front chassis. It happened twice. She had taken off her shoes. But the grip of naked feet also couldn't prevent the mistakes. She uttered a grimace after every painful noise, and became more rigid. She held the wheel firmly, drove cautiously, slowed down at every corner. The surge of the breeze through the window died whenever the car had to take a turn.

It wasn't very late, only half an hour past midnight. The blue car left the vibrant roads and entered the streets where the night was more obvious, with little sign of life. The poet had no idea where he was being driven to – the adventure was on.

Boredom didn't speak a word. She had to work very hard to make the small car follow her intent. Even the harmless FM radio seemed to distract her.

'Don't smoke. I cannot stand cigarettes,' said Boredom finally, when the poet was about to light a cigarette. She said the words almost angrily, in the tone of an order that invites defiance.

The poet didn't want to be stranded on a deserted street, in an unfamiliar part of the city. He reluctantly put back the cigarette in the pack, and felt a bit better when Boredom said, 'Thanks.'

The car moved towards the northern suburbs. A couple of zealous street dogs amused themselves by running after the car for a few frenetic moments. Half an hour later, the car approached a housing complex and came to a halt.

A row of four-storeyed houses made up the quiet neighbourhood. The feel of the night was very silent and very still – even the mild thump of closing doors caused a sudden embarrassment of creating too much noise.

The poet followed Boredom to the second floor. There was no elevator. The dusty staircase was narrow and well-lit; wall tiles depicting figures of divinity adorned the landing corners to ward off red tobacco-spit.

A surprise awaited the poet inside the living room. A full-grown German Shepherd greeted him with an indifference that relieved him. The dog was silent, well-trained and was used to seeing strangers at odd hours of the night. Boredom led the dog to a small balcony and bolted the door.

'Would you prefer a drink?' Boredom asked.

'No, thanks,' the poet said. 'That's good,' she replied, 'I don't like drunk men in bed.' Her voice and her attitude had no hint of spontaneity. The entire episode had a cold business-like air to it, a sense of non-monetary trading, a form of primitive barter system.

'You can put on music if you like. But not too loud. I will be back in a few minutes,' Boredom instructed. Then she entered a room and closed the door behind her.

The poet looked around. A metal frame containing a marriage photograph stood on a corner table. It depicted a smiling young Boredom and a bulky man with a heavy moustache. The picture was taken some years back. Boredom was no longer as slim as then.

At that very moment, the poet began to feel uncomfortable about the whole episode. The influence of alcohol had reduced itself to the advent of a headache.

He always carried a couple of analgesic tablets inside his wallet. He entered the kitchen and took out a bottle of water from the refrigerator. The tablet was smashed into pieces. He tore the wrapping carefully and gulped the powdered pieces, but he couldn't avoid tasting the bitterness of the medicine.

A moment later, he left the grubby unclean kitchen and came to the simplicity of the living room decorated with cane furniture and bright cushions. He surveyed the compact discs which were around the plain music system. Film songs, remixed versions of film songs and lounge instrumentals of film songs – none of them sparked his interest.

After a short while, Boredom returned. Underneath the loosely strapped cotton housecoat she was wearing nothing. The poet felt a stirring in his groin. He forgot about his headache. Boredom handed him a pack of dotted condoms.

'Open your shoes out there,' she ordered, 'and come to the bedroom.'

12

The erotic escapade of love without love – resembling the reality of animals, where there is only the truth of the present, no past or future, aroused an untamed energy within the poet. His desire stirred brutally without being tempered with any thought. He was absolved of all inhibitions and fears ... fear of questions, of being judged, of someone knocking on the door. He didn't even know the name of the woman; there was no nameplate on the entrance door. There was no obligation to recognize each other if they ever met again. There was nothing at stake, except for passion.

Freed of everything except his urge, the poet's lovemaking was fierce. And Boredom was one of those women who only let things be done to them. She was not keen to extract pleasure. That was the theme of her sexuality, to play the role of the weaker sex, to give in, to surrender.

After the climax, the poet lay on his back with a blank mind. His awareness had tasted a deep spacious darkness. It took a while for him to recover his breath. His eyes were still closed when he felt Boredom climbing over him to brush her breasts over his chest. That's the only initiative she took, and the second time round, the poet played the role, not that of a brute, but of a caring lover, more gentle and more affectionate.

13

'You cannot stay here for the night,' Boredom said, 'I will drive you to a taxi.' The poet looked at his watch, it was nearing 3 a.m.

He picked up his clothes from the floor and started putting them on. Then he noticed another photograph in the bedroom. Boredom was with a boy in a school uniform. The picture was taken in a hill station. The boy bore a strong resemblance to his mother.

It took ten minutes of driving around to find a taxi. The poet got out of Boredom's car without saying a word. Both of them avoided meeting each other's eyes. Boredom uttered a quick 'bye' and drove away, less cautiously than before.

The middle-aged taxi driver had intuited something. He smiled at the poet and asked for double the amount. The poet was being exploited, but he was in no mood to bargain. There was no other taxi in sight. He agreed, jumped into the backseat and lit a cigarette. He felt a deep sublime contentment.

The taxi sped away to the nightclub where his car was parked. He felt the cool air of the night against his face and the smoke stung his eyes.

A little while later, he felt a pleasant exhaustion. His fierce lovemaking had produced within him, with surprising intensity, the void of an orgasm. He still felt its effect – a buoyancy in his spine and a blankness of mind. No thoughts occurred to him. He felt emptied of himself. His arms felt light.

14

The Void of Orgasm

Modern minds of the ancient Indian philosophers provide a sublime interpretation of an orgasm.

Vijnanabhairava, also known as *Shaivopanishad*, an ancient text of the Shaiva Yoga system, has this verse, 'At the time of sexual intercourse with a woman, an absorption into her is brought about by excitement, and the final delight that ensues at orgasm betokens the delight of Brahman. This delight is (in reality) that of one's own Self.'

In tantric literature, the experience of innate bliss, with its divine power of purification, is regarded as essential to spiritual growth.

Hejavra Tantra declares 'erotic pleasure is not the real bliss' and goes to explain the levels of erotic pleasure leading to the bliss of an orgasm. 'First, ordinary joy, is the expectation of contact. The second, refined joy, is the desire for bliss. The third, the joy of cessation is from the destruction of passion and by this the fourth is experienced', and the fourth is the innate joy of bliss that doesn't arise from the elements of the body and is the feel of the imperishable supreme that abides in all living bodies, and from which 'existence, non-existence and anything whatsoever originate'.

Vimalaprabha or The Ornament of Stainless Light (a commentary on Kalachakra Tantra), also declares 'the soul has the size of the body', and 'the body is the abode of supreme imperishable bliss'. And an orgasm or mutable bliss momentarily reveals the immutable bliss of the void.

Hence, the void of orgasm isn't simply the climax of a sexual act; it is a wondrous experience of a non-dual mind *freed* of passion, craving, conceptualizations and desire.

In other words, an orgasm momentarily reveals the delight of a happy void, inherent and connatural, within the reality of the self.

15

Four days later, the marketing trainee called up the poet and fixed a meeting at a pub, at an odd hour of a Sunday afternoon. 'I have something important to tell you,' the marketing trainee announced. The tone of his words promised good news.

Half an hour later the poet reached the pub, played a game of pool with a middle-aged man and then occupied a table with a bottle of beer. He wondered about the good news, but he was certain that the news had something to do with a woman.

The poet wasn't wrong. The marketing trainee was always punctual; he arrived on time with a glowing face, and informed the poet that he was getting married next month.

The news surprised the poet. 'Isn't it a bit too early. You are just twenty-two. Are you sure?'

The marketing trainee said it was a 'family thing'; he had known the girl for years and his father wouldn't give him the money to go abroad unless he married an Indian girl from the same cultural background and social status.

'I have told you about my family. They like all new flashy things, but their mindset is very conservative. They are only modern from the outside,' the marketing trainee said in a defeated tone that lasted for an instant. 'But she is very nice, slim, fair and pretty; we have the same sun sign, our horoscopes have matched. She is also oral friendly.'

The poet smiled. 'What more can I say then,' he said, 'congratulations and good luck.'

16

One hour later, they had finished four bottles of beer and paid two visits each to the men's room.

'You remember what I used to tell you in high school?' the marketing trainee asked the poet. 'The only depth a man requires from a woman is six inches.'

Both of them smiled. The poet said he remembered. But he told the marketing trainee that his statement was coloured with ignorance, chauvinism and sexism. The marketing trainee became thoughtful. The reflective look didn't suit his face. He went silent and took a deep drag from his cigarette. 'Maybe you are right. But you know,' he contemplated, 'matters of the heart make things different.'

'So you think you love your fiancée?'

'Yes, I think. Perhaps, I don't know.'

The poet began to laugh. 'That's the most convincing reply I have ever heard in my life – yes, I think, perhaps, I don't know.'

The marketing trainee didn't see the humour. He was never comfortable with himself when he had to think deeply about something.

An argument erupted at the next table that was occupied by a young couple. The girl left her seat, and shouted, 'You never care to look into my soul,' and ran out of the pub. The embarrassed boyfriend got up sheepishly, kept some money on the table and walked out slowly, while calling someone on his cell phone.

Fifteen minutes later, the marketing trainee came up with a sudden proposal, 'I have a contact number. Only for foreign escorts. Mostly Russian, I think. Let's call them up. We can take them to your apartment!'

'No whores or holy men in my flat.'

The marketing trainee understood. He thought for a while – hotel rooms would make things more expensive, cheap ones cannot be trusted; they have hidden cameras. He thought of something else. 'I know of a nice brothel.

Not posh but very safe. They don't have AC rooms with laser lights and music but the girls are clean and good.'

'Aren't you getting married next month?'

'That's the point. I won't have time for fun after that.'

'Come on, man,' urged the marketing trainee, 'whores are tension-free, you don't have to worry whether they get an orgasm or not.'

The poet smiled. But he turned down the idea. 'No, I don't feel like it,' he said.

The marketing trainee looked displeased; he couldn't sell his idea to the poet, even with the help of alcohol.

The marketing trainee looked here and there for a while. Then he went back to the qualities of his fiancée which the poet had already heard three times.

After a while, unable to bear it any longer, the poet asked, 'How far is the brothel?'

17

It was nearly midnight; the poet was by the window smoking a cigarette slowly. He was thinking about the evening – his absurd experience at the brothel.

Now he understood why he had laughed at the absurd circumstance; he could have been nauseated, or felt the nothingness that preys on unguarded minds. But when he had grasped the ridiculousness of the situation at the brothel he had felt the pangs, not of shame, or of despair, but of laughter. The absurd situation didn't make him angry, didn't weaken him, didn't thrust him towards the lowest sense of himself, but freed him, and infused him with lightness.

It was a strange occurrence, but the poet wanted to understand what had caused the laughter.

He retreated from the window, lowered the volume of the music and lit another cigarette.

He remembered the moment. He was waiting for the prayer to get over and was looking around the cubicle. Then a realization occurred. He recalled that something in his self had come alive. A deeper sense illumined a wider perspective.

That's the insight he wanted.

He recalled that suddenly his condition had appeared to him as petty, absurd and ridiculous. He couldn't stop laughing at himself. With every pulsation of his stomach muscles, he had felt a freedom, an ascent, and a release.

Then he realized that he had fallen headlong into the mud of gloom after he returned home. He imagined unhappiness like a patch of soft mud. The more heaviness and coarse emotions one keeps on creating within oneself, the more one tends to sink deep into it, and manages to go nowhere. The inner self begins to be grabbed by narrow perspectives, and the broader perspective, rooted to the sense in the heart, with its consciousness being aware of one's own actions, tends to vanish. One gets into a loop of living carelessly without the guidance of sense and of meaning.

18

The poet stubbed his cigarette in the ashtray, went to the kitchen and poured himself a glass of orange juice.

He walked back through the living room. The lamp on the teakwood sideboard was still lit. The framed photographs of his mother looked at him. But for a change, the photographs didn't haunt him.

For the first time, after a long gap, the poet felt comfortable in the apartment. It felt like home.

But the apartment hadn't changed. The condition of his own self had transformed.

He was no longer restless and unhappy, his awareness had deepened.

Finally, he felt cleansed of the boredom, the despair and the dullness that had afflicted him, ever since he had returned home with a mind weakened with nostalgia and sentiments.

19

Next morning, the poet didn't wake up with heaviness. Sleep had refreshed him. As the day progressed, he felt the urge to write something. He sat at his desk and began to fill a notebook.

Then the phone rang. It was the marketing trainee. He wanted to go back to the brothel, and find the 'poet's girl'. But the offer didn't excite the poet. The marketing trainee tried his best to sell the idea. But this time around, it didn't work.

'Are you coming to the club over the weekend?' he finally asked. 'No,' the poet replied. 'I've got to do something else now. I will see you next month, at your marriage.'

The poet hung up, put the mobile on silent and went back to the desk. After a while, his mind recalled a boyhood memory.

20

It was during the time of his parents' divorce. He had to go to the court to tell the judge that he wanted to be with his mother. (The poet's elder brother, two years older than him, had already opted to live with his father.) The hearing was scheduled in the morning, but something had happened, and it was rescheduled. The poet had an hour to kill. His mother and his maternal grandfather were busy with the lawyers. He had nothing to do except drift through the corridors of the court.

After a while, he got bored. In a big room there was a large crowd. The poet slipped inside the room and sat down on the last bench to hear the proceedings.

A hearing had come to an end, after a few minutes, another one started. A name and a number were called out, a skinny young lawyer, not more than twenty-five years old, got up from the first row. A witness was called – a stern lady with a look of disgust – and she took the oath.

The poet had understood that the skinny lawyer was about to question the witness.

But the skinny lawyer took too long a time to frame each question. He kept on looking at a thick brief of papers which he was holding in his hands. Then he asked a question that the witness refused to answer. The judge informed that the first six paragraphs had already been cross-examined at the last hearing, six months back.

The skinny lawyer didn't have any other questions. He looked at his papers, while another lawyer asked him to hurry up. Impatience was rising in the court; there were a line of cases which had to be heard; there were many people and many lawyers.

Then the skinny lawyer finally asked the witness when the house pump had been purchased. The witness mentioned a date. Then he asked from where had the pump been purchased. The witness named a shop. Then, who had installed the pump, how long had it taken to install the pump, whether new pipe lines had to be laid, what was the thickness of the pipes, whether the pipes were ISO certified or locally made, whether the pipes had to be repaired, and so on and so forth.

The series of questions disgusted everyone – the witness, the judge, the opponent lawyer and others who were present in the court. Even the sixteen-year-old poet knew that the questions were irrelevant.

But the skinny lawyer refused to think of anything else other than the pipes. It was a ridiculous situation in the court, everyone was howling at the skinny lawyer with disgust and impatience. Then someone shouted: it was a case of defamation, between a landlady and her tenant, what had pipes got to do with it.

The poet started to laugh; the scene had developed into an absolute absurdity, a sham and a farce.

And then, something happened, the skinny lawyer began to confess before the judge. He said he knew nothing of the case. His senior had to go to the High Court to contest an urgent corporate case. He had shoved the brief in his hands and asked him to 'manage' and take another date. He had no time to read or to prepare. He had only read the first ten pages of the brief. So he could not carry on any further. Then he prayed for another date.

Everyone was aghast at the confession. The skinny lawyer was being ridiculed. But he wasn't ashamed. He was smiling! The poet at that time did not understand why the skinny lawyer felt so happy.

21

Sometimes, a truth of the past only becomes clear in the future. Now the poet understood the reason for the skinny lawyer's happiness.

The skinny lawyer could no longer play the absurd game; he wanted to free himself from the situation, from the meaninglessness of it; he wanted to be truthful.

And once he had made his confession, he had saved himself from the silly game that he was playing. Even though he was being ridiculed and humiliated, the aspect of his self that responds to truth was filled with joy and happiness.

That's why he was smiling. He had stopped himself from doing something that was dishonest, absurd and meaningless, and had chosen to do what was agreeable to meaning and soul-sense. And that had released him and had made him happy.

22

The next evening the poet was watching television. On one of the international channels, an astronaut was being interviewed. He was asked about his most memorable experience in space. The astronaut paused for a while and said: when he glimpsed the Earth from space, he realized how small the Earth was compared to the universe. If an asteroid destroyed the Earth, it would be an insignificant cosmic event, compared to the larger events which are taking place elsewhere: black holes are eating up planets and stars, galaxies are colliding with one another. Compared to such events, the destruction of the Earth would be like a small pebble that someone throws into a river; the news won't even make it to the galactic news channels.

Then the astronaut said that it had made him realize how precious life was. And now, whenever there was a brawl at the bar he could only view such actions as silly absurdities.

He said, if everyone went to space then their perspectives would change forever – people would realize the broader picture, learn to honour life, protect the planet and stop killing and exploiting each other.

'This us and them mindset makes me mad,' the astronaut continued. 'It's one single planet with seven billion human beings. There is no us and them. It's a delusion and an error.'

The poet had found his insight. Absurdity is created when one looks at pettiness from a broader perspective. It's an inner realization that can discriminate between pettiness and generosity, stupidities and wisdom, myopia and vision.

Then the astronaut, who had now retired, was asked, how he spent his time after retirement.

'Within my daily life,' he replied, 'I only do the good, the harmless and the better.'

The words inspired the poet. Now it was time to organize his life, to set goals and to think about the future.

23

The apartment had been lying locked for nearly two years – the upholstery needed to be changed, the curtains needed a trip to the dry-cleaners, the kitchen and the bathrooms needed new fittings, the walls needed a coat of paint and the mahogany furniture needed polish.

For the next two weeks, the poet set out to renew the look of the apartment. He pondered over shades of colours, took quotations, negotiated rates and then exhausted himself with the painters and the polishers – who tried their best to encroach upon each other's territories – and the poet had to keep everyone happy so that they could do their work properly without needless delays caused by silly arguments.

The poet didn't like heavy mahogany furniture. The daybed, the sofa sets, the sideboards, the corner tables, the wall mirrors, the marble consoles, all made the living room look too rich for his taste. But he knew they had been precious to his mother. He remembered the relief on his mother's face when the relocation company managed to move everything without causing any damage. Burdened by nostalgia, the poet couldn't transform the living space into something warmer and simple.

But the third bedroom was vacant; there was only a large antique almirah that he shifted to the master bedroom and converted the space into a lounge-cum-study. He didn't think of hiring carpenters and burden himself with plywood, glue, nails and hidden costs. He visited various furniture shops and bought tables, lamps, chairs and book shelves. He also bought a toolkit, drilled holes in the walls, hammered wood pallets, fixed the screws and hung the shelves.

After all this, he arranged the furniture, shifted the books from his bedroom and took special care to get the lighting right.

The lounge-cum-study was finally done. Even though he had his laptop, he felt the need for a desktop computer with good speakers, new stationery and a broadband connection.

The poet got the broadband connection within a day. Then he bought a branded all-in-one desktop computer, speakers and a multifunctional printer.

Then he arranged all the important documents – related to himself, the apartment, bank, car, insurance, parents' divorce – into well-marked folders, and stored them in a file cabinet.

After all this, he felt better, he knew what was where. The apartment looked nice and neat, and it felt calm, peaceful and fresh.

24

The poet soon regained his inner life; he began to read, write, daydream. He entered into his usual habit of contemplating about philosophical and spiritual ideas that pleased him deeply and brought about an inner calmness.

At this time, he received a Facebook message from a college senior who used to be the editor of the literary magazine to which the poet contributed regularly.

The senior informed him that he had founded a website called *India Youth Culture Post* – an online culture magazine that would be publishing short fiction, essays, poetry, art, photography, independent music videos, short films and news articles related to social activism of various kinds.

The senior asked the poet whether he wished to be a part of the editorial team and do eight to ten hours of online work per week without any pay. The poet was delighted with the news and called his senior immediately.

The senior explained that the venture was being funded by friend-sourcing and had no revenue model. 'But it has purpose,' the poet remarked, 'and that is more important.' He agreed to join the team.

'The old fogeys are standing in the way of real progress. The space for serious cultural expression is shrinking. Everything is being viewed with the eyes of commerce and entertainment. Youth is being exploited as a market to sell products to, and a gradual dumbing down of society is happening,' the senior told the poet.

'Our monthly online magazine will counter all this through creative expression. Social media will help us connect with like-minded people and form dedicated groups for different spheres and issues.'

The poet loved the idea and suggested a couple of other friends who could be of help.

'Yes, get them involved,' the senior replied, 'but you will also have to contribute with your writings.'

'I can contribute poems and essays,' the poet said.

'The poetry section of the inaugural edition is full,' the senior replied, 'You can submit for the later editions, of course. But I want to commission an essay for the launch edition.'

'Do you have a topic in mind?' asked the poet.

'Yes, I have. Remember once we had a discussion about the concept of zero. So will you write about the history of zero? I think that would be interesting.'

The poet thought for a moment and said, 'Zero denotes the void. And it's a myriad void. The essay could discuss the various ideas of the void from various perspectives.'

'Do what you please,' the senior replied. 'You can change the title as well, as long the history of zero is written about. The word limit is two thousand. You have two weeks.'

25

When the poet had turned thirteen, he had received a two feet telescope for his birthday. He used to take his telescope to the terrace of his house and watch the night sky.

The haze over the city prevented him from seeing all the planets and the constellations, but he soon developed a love for the night sky.

Whenever he would gaze at the stars and imagine the depth of the universe, the sense of search and the feel of wonder would fill his heart.

He had read enough about the universe to know that there exists a point in space where every orb of light is not a star but a galaxy.

'There are billions of galaxies in the universe,' he would tell his elder brother, who never seemed to be emotionally moved by this wondrous idea that 'there could be billions of universes in the great void of the cosmos'.

Ever since that time, the poet had developed a strange, deep fondness for the word 'void', and over the years he had found many more meanings associated with that word.

When the opportunity arrived to write an essay about the void, he was filled with the same boyhood enthusiasm of gazing at the night sky with a telescope.

The poet referred to his old notebooks where he had jotted down many ideas of the void over the years, did an extensive research through the nights and woke up at noon for the next ten days.

Then he wrote the essay in four days and submitted it, just an hour before the deadline.

The essay delighted his senior and it was published in the inaugural online edition of *India Youth Culture Post*.

26

The History of Void

The Unhappy Void

The void of *nothingness* is the prime theme of existentialist philosophy. Kierkegaard indicated, the subjective fear of 'nothing' preys underneath all human experience. Sartre concluded consciousness is nothingness. Heidegger wrote, 'nothing is more original than the not and negation'. But he also discovered a link between the sense of *nothingness* and boredom – 'Profound boredom, drifting here and there in the abysses of our existence like a muffling fog, removes all things and human beings and oneself ... into a remarkable indifference.'

The void, in philosophy, is *nothingness* – a condition of the mind afflicted with boredom, anxiety, nausea, absurdity and alienation.

Those who suffer from *nothingness* often commit the most desperate of all human actions. They damage themselves, their lives and their relations.

The Happy Void

Maitri Upanishad describes the innermost self as 'pure and purified, void and tranquil'. The *shunya* or *akash* or the void is ever present as the eternal background to the self.

Self-realization ends with the realization of the void – a spiritual space-like state that experiences the pure self without attributes.

The practice of meditation plays the central role in the understanding of the void. In the Buddhist canon, monk Nagasena recapitulates for King Menandros the twenty-eight positive advantages of solitary meditation. In *Svetasvatara Upanishad* where the mind is compared to vicious steeds which require reigns of silence, meditation is viewed as a raft 'to cross all the rivers of life so fraught with peril'.

Unlike the void of philosophy – the negative *nothingness* – the void of spirituality is the positive *nothing* – the origin of merits (love, truth, bliss and understanding) and divine essence.

The Unknown Void

Einstein introduced the term 'cosmological constant' to describe an independent force holding together the web of the universe, balancing the forces of gravity. Einstein called this concept his biggest blunder, but now advancements in physics have proved that Einstein was, in fact, *right*. Empty space is pervaded with 'cosmological constant', or in the words of *Surangama Sutra*, where Sakyamuni Buddha converses with Ananda, 'open space is not nothingness'.

Scientists also declare, the vacuum energy of empty space is enormous, and the all-encompassing energy field might be positive or negative. Elementary particles can simply pop out of this void, or in the words of the ancient Taoist manifesto, *Tao Te Ching* by Lao Tzu, 'Something is produced from Nothing!'

In the seventeenth century, the metaphysical poet John Donne lamented the propagation of the mechanistic idea of the universe by science that had robbed 'wise nature' of beauty and mystery. But now the wheel has turned full circle. 'Wise nature' has regained her honour. The mysterious void – full of 'unseen dark energy' – has almost transformed twenty-first century scientists into mystical philosophers who are just beginning to discover its secrets and magic.

Ironically, in the twentieth century, Heidegger complained 'science wants to know nothing of the nothing', but now in the twenty-first century, science wants to understand that 'nothing'. And great efforts are being made in tunnels under Geneva to discover the first of the five Higgs Bosons and also understand the workings of the invisible void.

Whatever term may be used to describe the void, *Mahashunya* or *cosmological constant*, whatever way one might try to understand it – through spirituality or through science – the truth remains that the void is real, and it exists. Ancient mysticism has now become scientific truths, and the perceived rivals – science and spirituality – are now playing for the same team.

It must also be pointed out that the number that has been vital for scientific progress is 0 (zero) – without which mathematics, hence the technological progress of mankind, wouldn't have been possible; the language the computer understands is still binary – 0 and 1.

Behind all the tumult of life, the void of 0 continues to play its crucial fundamental role.

The History of Zero

The history of zero begins in ancient India. The number was the symbolic representation of the spiritual concept of the *shunya* or the void.

The mathematician Pingala (fifth-third century BCE) devised a binary system. But he didn't use the symbol 0, instead he used the Sanskrit *shunya* to refer to the concept of void.

In 500 CE, Aryabhatta devised a number system and used the word *kha* to denote emptiness. Use of the *bindu* or the dot to denote the void was also common in earlier Sankrit texts.

In 628 CE, Brahmagupta in his classic work *Brahmasphutasiddhanta*, or The Opening of the Universe, was the first mathematician in human history to extend arithmetic into negatives and zero. But he too didn't create the symbol 0. An Indian stone inscription, referring to the town of Gwalior, dating to 876 CE is the earliest authentic record of zero as 0. It denotes the numbers 270 and 50, and 0 appears, as it appears now.

So between 628 CE and 876 CE, an unknown Indian genius, pondering with numbers, had a sudden insight to create the numerical symbol 0 to denote the concept of zero. He created 0, completely unaware what crucial role 0 will play in the development of mathematics, of science, of technology and of civilizations.

Trading links across the Arabian Sea flourished in the eleventh century. Loaded with aromatic spices and other riches, one unknown ship carried the Indian numerical system to the Middle East.

Globalization of 0 had begun. Al-Khwarizmi wrote *Al-khwarizmi on the Hindu Art of Reckoning* that describes the Indian system of numerals based on all the ten numbers – 1 to 9 – and 0. Ibn Ezra and Al-Samawal contributed other notable works.

The Hindu-Arabic numerical system reached Europe via Andalusia in Spain in the late eleventh century.

In 1200 CE, Italian mathematician Fibonacci described the Indian number system in his work *Liber Abaci*, and introduced 0 in Europe.

What had begun as the Sanskrit *shunya* became *sifr* (meaning empty or vacant) in Arabic. Then it became zephyr or zephyrum in Latin. Fibonacci turned zephyrum into zefiro in Italian. And zefiro was concised to zero in the Venetian dialect.

By the thirteenth century, the Indian number system had reached China, and the Chinese mathematicians had started to use the symbol 0. In 1247 CE, Ch'in Chiu-Shao wrote *Mathematical Treatise in Nine Sections* and in 1303 CE, Zhu Shijie wrote *Jade Mirror of the Four Elements*.

By 1600s, 0 came into widespread use, just in time for the development of complex mathematics, of science, of technology and the roots of modernity.

The unlikely hero behind all such progress was the all-powerful zero.

Born in Hindu-Buddhist India, spread by the Islamic Arabs and put to use to develop mathematics by the Christian Europeans, the void of 0 has come a long way to become the language of super computers.

Its story is like a parable that unifies all the major religions, and the regions of our earth – from the East to the West, via the Middle East.

*

The history of 0 will remain incomplete if one doesn't mention the origin of the void concept.

The concept of the spiritual *shunya* or the inner self – the void within – was first developed in the *Upanishads* which are dated from 2000 BCE to 500 BCE.

But the concept of the cosmological *Mahashunya* or the Great Void had arrived more than four thousand years ago.

The poem – *In the beginning* – from the *Rig Veda*, expounds the evolution of the universe, of gods and all creatures from the primal darkness of the cosmic void.

At the end of the verse, a doubt is raised whether God created the void or whether he did not. It ends with this astonishing line: 'Only he ... in the highest heaven knows, or perhaps, he doesn't know!'

A few thousand years later – when the effort to understand the void had become a prime theme in all spiritual philosophies – the Indian mathematician Bhaskara had the inspiration to compare the special properties of zero to that of God – 'no change takes place in the infinite and immutable God when worlds are created or destroyed, though numerous orders of beings are absorbed or put forth'.

Bhaskara linked the zero with the cosmic void of the *Rig Veda* and wrote down $0^2 = 0$ and $\sqrt{0} = 0$ while thinking about the properties of the unalterable Omniscience.

The origin of the mathematical 0 was inspired by the spiritual *shunya*.

And the *shunya*, from the cosmic *Mahashunya* or the Great-Void, the *tajjalan* – 'from which all things are born, into which they dissolve, and in which they live and move.'

0 is not merely a mathematical number. 0 signifies the void – within the self and within the cosmos – the reality of 'nothing', that *is* 'something'.

*

It was Einstein's idea of the 'cosmological constant' that is now termed as 'dark energy' – a term that was coined by Einstein himself. He said properties of this new energy form were either nothing, or it was a 'non-observable negative density in interstellar spaces' and 'that turns out to be dark energy'.

More than 4000 years before Einstein, sage Yajnavalka had a conversation with Gargi Vacaknavi. The dialogue is documented in the 'Aranyaka' or 'the forest book', that is now known as *Brihadaranyaka Upanishad*, the very first of the many Upanishads.

Gargi asked Yajnavalka, 'that which is above the sky, which is below the earth, which is between sky and earth – that which men speak of as past, present and future; on what is *that* woven.'

Yajnavalka answered her, 'That is woven on space.'

'On what then is space woven?' Gargi asked.

Yajnavalka replied, '... that is what is called *Imperishable*.'

In other words, Yajnavalka revealed that time is woven on space, and space is woven on the *Imperishable*.

Or, in words of Yajnavalka and Einstein, 'space-time' is woven on 'dark energy' whose 'properties' Yajnavalka described in four poetic paragraphs, and used phrases like 'it is not darkness, nor light', 'it has no within nor without', 'it has no face nor measure', 'the unseen seer, the unheard hearer, the un-thought thinker, the un-understood understander'.

Hence, Yajnavalka had gone beyond 'space-time' and identified the modern 'dark energy' as the 'imperishable', whose symbol is the void of 0.

In the twenty-first century, when one recalls the long history of zero, one gets infused with a strange sense of wonder.

All the progress of mathematics, then of science, of technology and of modernity was rooted from 0 – an abstract spiritual concept to denote a reality that is invisible, mysterious, divine and unknown.

The reality of 'dark energy' or 'the imperishable' is not only present in the far reaches of space, but is also ever present within the reality of the human self.

27

The poet didn't realize that while he wrote the essay, the essay was about to write a significant part of his future.

The last paragraph of his essay had a profound effect on him. He felt something change within. He thought – what he had understood intellectually, he had to realize through experience.

During his research for the essay, he had encountered a certain line from *Hevajra Tantra* that had mystified him:

'Great knowledge located in all bodies is non-dual as well as dual in nature.'

The poet felt that a part of him understood the line, but the understanding wasn't really quite clear.

At that moment he made a resolution – 'I have to develop greater awareness because I am trying to see the unseen. I have to make an effort to answer the question that pursues me. What is the human self?'

If there is a question, then there is an answer, whether one knows it or not.

The poet didn't know the answer yet, and the quest for it became the leitmotif of his inner life. And he became more determined to make the invisible visible through his understanding.

At that very moment he was filled with the same boyhood enthusiasm of gazing at the night sky with a telescope.

He was filled to the core with the same wonder and search.

28

Two months later the poet was visited by his maternal grandparents. They lived in the city of his birth and where he had spent his boyhood, the city he had left five years ago. (The poet's paternal grandparents had died when his father was a boy.)

The poet had graduated in Economics with a first class; he had also topped the university in two papers.

The certificate and the marksheet arrived through post. The poet got them laminated. It was a good idea – the lamination also warded off tears.

The poet's grandmother, with moist eyes, took the graduation certificate in her hand and walked to a photograph of the poet's mother and started to cry. 'Your mother would have been very proud,' she told the poet.

His grandfather was less sentimental, but he reminded the poet what his mother wanted him to do. 'I know,' the poet replied, 'she was competing with father. He had sent my brother overseas. She also wanted to do that. But, as you know, dreams cannot be thrust upon someone.'

'That's right,' his grandfather said, 'you don't have to do what you don't like. But have you thought about anything else?'

'I have,' the poet replied. 'A lot has happened already.'

During college he had written several youth-related articles for a newspaper. On that basis, he had applied for a job in a leading English daily. He had been called for an interview, and had been offered a job. The training was to begin in four weeks. He would specialize in feature writing. The salary wasn't a big issue – the interest from the term deposits was enough to take care of the bills.

'Journalism!' the grandfather exclaimed, and went into a spell of thought. He leaned back on the sofa, took slow puffs from his pipe and made some rapid calculations.

The poet's grandfather was a respected doctor, a general physician, with a passion for astrology. One of his grandfather's elder brothers was an astrologer and a spiritualist – the poet's great grandfather had bestowed on him the title of the family's black sheep. He was twenty-six when he left home in favour of a spiritual life, leaving behind many books on astrology, which were lying in neglect till the poet's grandfather rescued them from rusted trunks tucked away within the chaos of a storeroom – full of broken sentimental things which could not be displayed or thrown away.

Since the discovery of the books, the poet's grandfather had continued to develop his personal hobby along with his medical practice. No one minded it, till a day arrived when he wrote in a prescription – meant to cure a chronic liver ailment – along with two names of medicines, 'an 8-carat, yellow sapphire to be worn on the index finger on a Thursday morning to appease Jupiter'. The news of the prescription spread fast and a vernacular newspaper published it on the front page. He had to give an explanation to the medical council. But he refused to denounce astrology and only apologized for 'deviating from the precepts of a doctor'. He didn't care if the council revoked his license, but he was let off with a warning. The controversy didn't ruin his reputation, instead more people started calling on him to take advice from the 'doctor-astrologer'. Such an arrangement tired the poet's grandfather; he retired from the medical profession and hung up his stethoscope. His circumstances changed for the better; there was no need to earn money to keep up with the rising cost of living; he had bought two acres of land many years ago; sometime back he benefitted from the construction boom in the city and struck a deal with a property developer to construct a residential complex; he received a large sum of money that he had already invested in post office savings and bank fixed deposits. He was also tired of dealing with so many diseases and wished to spend his time nurturing his hobby to something more satisfying. But he didn't want to offer his astrological services to 'confused and superstitious people', and immerse himself in all kinds of terrible problems, silly questions and wishful fantasies. So he devised a way to keep alive his interest and also contribute from a distance; he started writing for various astrological magazines, made a new group of friends and started a new life at sixty.

The poet waited for his grandfather to react to his decision to pursue journalism.

His grandfather reflected deeply and took puffs from his pipe.

'Tenth lord Mercury in the eleventh, with Sun, with the aspect of Jupiter and Moon. Quite appropriate,' he said.

29

For the next couple of days, the poet drove his grandparents around the city, visited markets and shopping malls, where the poet discovered his grandmother's terror of escalators and his grandfather's love of time. He visited all the shops selling timepieces, and finally bought the poet an expensive watch to mark his graduation.

Later that night, after eating the delicious dinner prepared by his grandmother, his grandfather took him aside for a conversation.

He needed the poet to accompany him to a lawyer to take care of the legal succession certificate and other formalities, and also for some bank work that had been pending for a long time. 'We are old now, anything can happen any day,' he said. 'I have made a will and have brought a draft that I want to deposit in your bank as a joint fixed deposit.' Then he said, 'There is also some of your grandmother's jewellery that you have to put in the bank locker.'

Then grandfather took a deep breath, relaxed himself on the sofa, took another puff from his pipe and said, 'I know you dislike your father. You think he is the cause of your mother's death. If only he had lived his life more responsibly then the divorce wouldn't have happened, neither perhaps, all those unfortunate things, which ended with your mother's death.'

'The truth is,' his grandfather said slowly, in a tone of deep reflection, 'I also feel like that. And like you, I have also never liked your father.'

'But as you know, in life one has to accept many things and move on,' he continued. 'Your father is a lonely man now. Your brother is abroad, he will never come back. And that woman has also left him. I also hear he had a prostrate operation.'

'That's not a surprise,' the poet said, 'all your unfair actions return as troubles and problems.'

'Yes, that's right, one has to bear the consequences of one's deeds. You cannot escape that, as you cannot escape death,' his grandfather said, 'but you also have your own responsibility, after all, he is your father.'

'What are you hinting at?' the poet asked firmly. 'I should go back to him and say how much I love him?'

'No, not at all,' replied his grandfather, 'I don't mean that. All I want to say is, sometimes one has to act, not out of love, but out of responsibility.'

'What irresponsible thing have I done to him?'

'No, you haven't done anything to him. But accept his calls for a change. Hear what he has to say. Don't disconnect the phone when he calls you.'

'Who told you I do that?'

'Your father called me up. He thinks I poison your mind against him. He accused me of taking his own son away from him.'

The poet gave a sneer, 'He was always stupid, it's no surprise that he thinks that way.'

His grandfather wanted to say something but the poet continued, 'Let's be clear, grandfather, I don't want him in my life. I know he has enough money. He has his debauched friends. He has his club. He has his whisky. But I cannot relate to anything that he stands for – his way of thinking. He cannot see anything beyond money, pleasure and power. He represents ego-driven corrupt people who do more harm than good in our world. I have my right to ignore someone whom I cannot respect. But I am not after any revenge, I am not out to harm him. Truth is he means nothing to me. I am happy with this void. And I want to keep it that way.'

30

When the poet was twelve-years-old he had gone to a five-star hotel along with his family, like they often did, to have dinner.

The poet's father didn't want to have any dessert, so three chocolate desserts were ordered – one each for the poet's mother, the poet's elder brother and the poet. Then the poet's father changed his mind, one more chocolate dessert was added to the order.

After they had finished the poet's father asked for the 'check'.

The poet was always entrusted with the responsibility of checking the bill. He discovered that the fourth chocolate dessert, consumed by his father, hadn't been included in the amount. He told his father about it. His father said it was alright, it didn't matter. He quickly paid the amount and rushed his family out of the restaurant.

The poet's mother understood what had happened only when they were in the car.

The poet's father reasoned and his brother agreed that it was the responsibility of the hotel staff to bill correctly and if they hadn't done their job properly it was their bad luck, they were to be blamed.

The poet never understood such reasoning, neither did his mother.

When they reached home, there was another fight between his parents. Many a glassware broke that night.

Something also broke within the poet.

His father had sold his honour and respect for a free chocolate dessert!

From that night, the poet couldn't get himself to respect his father. He appeared to him as a selfish opportunist, devoid of principles and ethics.

When a person cannot respect another person, it doesn't mean one has to show disrespect.

The poet never behaved badly with his father, but the image of his father continued to fall in his eyes, and never recovered.

31

After his grandparents left, the poet, one evening, was watching one of the travel channels which was airing a programme on Benares – one of the oldest living cities.

The poet had been to Benares twice – his father had inherited a house from his uncle there. The programme rekindled many memories.

His mother had also told the poet that he was conceived when his parents were visiting Benares.

The next day, he went to a bookshop and bought a book on Benares. He had two weeks to go before he started his training. There was enough time for him to do some travelling.

A visit to the railway booking website got him a ticket. It would take an hour to pack his backpack. The train was scheduled to leave that very evening.

32

In the high-rise building where the poet lived, each floor had three apartments – one had four bedrooms, one three and the last one had one. The one-bedroom apartment adjacent to the poet's home had been sealed on the orders of the court, and it would remain so till a pending dispute over its ownership was lawfully resolved.

In the largest apartment lived a lady of motherly disposition. She was a nice person – kind and religious – with a delightful sense of morality. She refused to light any incense stick with a lighter. The lighter was immoral – it was designed to light cigarettes – only a matchbox was pure enough to light incense sticks meant for worship. The lady's husband was the senior vice president of an industrial firm. He was a grim man with a passion for bridge – he travelled frequently to take part in various international competitions. They were childless.

The lady was more accustomed to the absence of her husband, than his presence. She had no friends, no independent life. Religious and social obligations filled her time. Not a single religious occasion, however minor it might be, went unnoticed by her.

She also concerned herself with the wide extension of the great Indian family. The person married to the first cousin-sister's third daughter, from the mother's side, also meant a lot to her.

She surrounded herself with a fleet of relatives, who were always eager to lounge in her apartment, and enjoy her favours and hospitality.

The lady had been very helpful to the poet and his mother. Now she viewed him as the son she never had. But she was wise enough not to encroach upon the poet's personal space. She never tried to fill the void left behind by his mother.

During the poet's absence (his two remaining years at college after his mother's death) she had arranged to maintain the car – one of her drivers gave it a regular start and took it out for fitness drives – and also the flat, which was cleaned every weekend, the yearly corporation tax was also deposited, so was the electricity bill till the poet's grandfather made other arrangements.

The poet's grandfather always thanked her for what she did for the poet. She always said of the poet, 'He is like my son.'

The poet returned her favours by asking for her advice on mundane matters which didn't require her involvement. Whenever he went to her for something, her face lit up with happiness.

She engaged a part-time cook and a cleaning lady, when the poet refused to keep anyone in his flat on a full-time basis. She was also overanxious with the poet's youthful status. She rejected many young women, and finally employed two mature ladies, with children and family.

She had an army of housestaff, one of whom would often go over to the poet's apartment to deliver food in sparkling white casseroles. When the poet would go to return the casseroles he would sit for a while and chat with her. She would recall the pristine hill station where she spent the first sixteen years of her life. She would also speak of a beautiful niece of hers, possessing countless qualities, who was studying to become a software professional. And on one occasion she embarrassed the poet by inviting him to her apartment when her niece was also present. To avoid such harassment in future, the poet, during the conversation, referred to the sweetheart as his girlfriend who had gone to Germany and would return after a few years. It worked. The poet didn't see the niece ever again.

33

Before leaving for Benares, the poet left the keys with the lady. He had to tell her where he was going. The news pleased her; she gave him a hundred and one rupees. The poet had to buy offerings for the deity – in the most famous temple – and bring back flowers, *bilva* leaves and the thick brown circular sweets.

She told the poet to be careful and to give her a call if he needed anything. 'I have a relative, a very nice person, a chartered accountant. His son is in the navy. His friend lives in Benares. If you call me tomorrow, I can get his phone number and address.'

'Thank you for your concern,' the poet said, 'but there is no real need to take such trouble. I can take care of myself.'

As the poet entered the elevator, she called out, without saying the poet's name as that would bring ill omen to him, 'Listen, don't eat anything from strangers.'

The poet could only smile. Her naïve simplicity touched him. She was no longer just an ideal neighbour – helpful and non-inquisitive – but had taken upon the role of a motherly aunt that pleased her excessively, and that in turn, pleased the poet.

34

In the compartment of the train, the poet's companions were a noisy group of eunuchs.

The poet remembered a superstition – to glimpse a eunuch before a journey was considered to be a good omen.

He avoided looking at them, plugged in his MP3 player and stared out of the window.

After an hour, he ate a cheese sandwich, drank water, used his backpack as a pillow, stretched out on his berth, and fell asleep.

35

The poet woke up to encounter a startling stillness. The train had halted at a small station. It seemed, the train had died.

It was deep into the night. Everyone else in the compartment was fast asleep. A blue night lamp was on.

Through the window he couldn't glimpse any movement. The platform was deserted. There was hardly any sound.

The poet tried to read the name of the station. He saw a yellow-black sign and read the name. Then he heard a whistle; the train jerked twice, and began to move.

Through dark fields, the train picked up speed. There was not a speck of light anywhere in the heart of distance. But the slice of moon was awake. The sky had stars.

The poet was gripped by a childlike emotion. He wondered: how would a classical poet describe the slice of the moon. He probably would say: *the ferry boat of dark skies that sails to distant shores.*

Spurts of innocence save a man from many monsters. The line made the poet happy. He smiled, withdrew from the window, and fell on the backpack.

The moving train worked like a cradle. He sank slowly, into the deep void within his self – the void of dissolution, of mystery, of renewal – another manifestation of the myriad void – the happy void of repose, of oblivion, of sleep.

Circles and Spheres

1

The frayed wooden planks creaked with every movement. The bamboo oars were motionless in the gurgling waters. The predawn light was changing its colour, the farthest stars were vanishing from the sky. From the shore, the river seemed to be flowing with the serenity of a wise sage. But it still required the energy and the experience of two oarsmen – one young and one old – to row the boat upstream and then allow the downstream flow to carry the boat like the graceful swim of a swan.

Kashi, the old part of Benares, is situated where the river takes a crescent-shaped turn towards the north, waits expectantly on the western banks, gazing towards the east, for the first glimpse of the morning sun.

There were many other boats; they too were silent and expectant, waiting patiently, for the drama to erupt.

Then it happened. Light struck sound. The first rays broke through. And at that very moment, hundreds of conches, bells and drums exploded with the light.

2

On his arrival, the poet wanted to find an accommodation by the river, within the circle of the old city of Kashi.

At the railway station he bought a guide meant for tourists. He went through the book at a restaurant while he ate his breakfast. Then for the sake of some local knowledge, he decided to speak to a middle-aged local who was sitting next to his table. The man gave his inputs, while the poet finished his milky tea. He thanked the man, paid for his breakfast, grabbed his backpack and came out to the street. He hopped onto a cycle rickshaw and told the rickshaw puller to take him to the old city. The rickshaw puller told him, there was no place to stay in the old city and suggested the name of a new hotel in the congested new city area, which would be good for a young man travelling alone.

'You won't get any commission from that hotel,' the poet told him. 'Take me where I told you to. Don't try to mislead people out of selfish interests. It's not my first time in Benares. I have been here before.'

3

After an hour's search, the poet found himself in front of a building. At the entrance, a smelly ram was tied to a pole, amidst a splash of brown grass. The tiny dark reception space, under the steep staircase, contained a desk and a small man on a chair.

'I have a large room with a balcony that faces the river. It's on the third floor. But it has no attached bath. You have to share the common facilities.'

'That's not a problem,' the poet said. He didn't expect the old building, polished with age, to offer modern utilities.

'Do you want to see the room?'
'No. It's alright.'

'How long will you stay?' the man asked and produced a tattered register from a drawer of the desk.

'I don't know, as long I like it, I suppose.'

'Should I sign you in for a month?'

The poet smiled. 'I don't have a month to spare. Maybe a week, that's all.'

The man asked for an advance. The poet gave the money, signed the register, exchanged a few words, took the key and headed to his room.

4

The room wasn't morbid as the poet had expected it to be. It was large and clean. It had a double bed on the oxide floor, a desk, a chair and a faded mirror.

The poet dumped his backpack on the mattress, switched on the ceiling fan and went to the small balcony.

The balcony offered a splendid view of the river. His body was tired but his mind was lively.

He observed the boats loaded with pilgrims and tourists, and the terraces of other old buildings, where there were breezy restaurants with wrought iron furniture and large colourful umbrellas.

The nearness of flowing water affected him like rain. A serene gravity touched him. Time slowed down with the deepening of his breath. A clear awareness, forever wakeful, at the background of his self, infused his mind.

Then he looked at the horizon. The other shore had merged with the landscape of sparse vegetation, brown-ashen in hue, and had converged with the sky.

It was a beautiful day. The water was cool, the distance was deep. Everything was very peaceful. The odour of silence was very old, and very ancient.

5

For the next few days, the poet explored Kashi with the slow ease created by the reflective and healing effect of solitary travel.

Centuries of worship had added a sort of deep flavour to the air of the old city. In spite of the congestion and the narrow lanes, the poet could feel a sense of space within himself.

Soon he discovered quiet spots by the river and charming little restaurants where he had his meals.

The poet also remembered the second time he had visited the old city with his family. The poet had wandered away to explore the lanes, his mother had to search for him, and he had received one tight slap as a reward for his curiosity. But the memory of the slap now made him smile. He remembered the peach-coloured house where they had stayed; it was located in the new part of Benares, away from the riverside. It was a big house with a square courtyard, large doors, high rooms with dark wooden beams and strange electrical switches, which his mother had fiercely warned him not to touch. There was also a lawn, uncared for, with four large trees at the back, where the poet had played cricket with his brother. There was a major argument between his parents. His father had wanted to sell the property that he had inherited from his uncle. But his mother did not agree. She felt that there was no need for money and the house could be put to some better use. But his father had already made a secret deal with a businessman who wanted to develop the land and construct a hotel. The argument lead to a fight when the power suddenly went out. Candles had to be lighted. Then the stout businessman arrived with two poker-faced companions. He paid more in cash than the amount scribbled on the cheque. The poet's father signed the papers under the glare of torch lights.

And now, years had passed, there was no house and no family. But the thought didn't fill the poet with nostalgia or despair; he didn't have to counsel himself to raise his spirit. One has to accept reality and learn to move on, without being burdened by it.

The poet brought back his awareness to the moment. He was in one of the lanes, walking past a sweet shop. He heard the sizzle of delicacies emanating from the cauldron of boiling oil. The aroma followed him, till he took a detour through a tiny lane between old walls with antiquated textures, and approached the river.

6

The poet went down to the steps, found a place to sit, lit a cigarette and watched the river.

Lights of the city danced on the waters. The moon was waxing in the sky.

The poet felt a mysterious love for the old city. In spite of the chaos and the clamour, the deep awareness of peaceful solitude never deserted him.

Wherever one may go one can only experience oneself.

Every place has its own sphere with its own ingredients, characteristics and flavour. People react differently to different places as they react differently to people, things situations and circumstances.

This happens, the poet thought, because every individual self with its distinct characteristics and an array of inclinations, tends to circle and repeat experiences within different spheres.

That's why he liked the *sphere* of the old city, because the *theme* of his self responded to its *sphere*.

7

The poet was back in his room. The fluorescent tube had a fault. He had reported the problem to the man at the reception. But nothing had been done about it. He didn't want to strain his eyes under the dim 40-watt bulb that emitted a red light. He stood on the chair and tried to readjust the tube. The tube flickered with clicking noises, and then, when the hope of the poet was about vanish, it lit up.

The poet changed into his shorts and stretched out on the bed. He had bought a couple books from a tiny bookshop. He read a few pages, then went back to his new notebook, and began to fill it with poetry and reflections.

The poet thought: it may seem that an individual passes through the course of life like the waters of a river, originating, passing and dissolving in a set course of birth, life and death.

But an individual simply doesn't pass through various experiences but also through various *spheres* of experiences.

Now that he had chosen to be a journalist he would enter into a sphere with its own distinct ingredients.

And in life there are countless such *spheres* related to work, location, activity and interest.

A strange-looking insect came into the room from the open door of the balcony. It hit the tube light, whirled around and flew back into the night.

The poet got up from the desk, drank water and went back to his notebook. Then he wrote:

> Self is a theme,
> a pattern,
> an array of traits
> responding to time.

8

Every morning in Kashi, a frenzy of temple sounds greet the rising sun. The loud resonance emerges from the deep interiors of the shore temples and echoes all over the old city till the structures along the banks – a jumble of stone buildings with dark watermarks, of antiquated paleness, of an ashen milieu – and hundreds of early morning bathers get soaked and drenched in the brazen hue of dawn.

The series of stone steps leading to the river gets populated with the faithful who bathe in the Ganges and chant the Gayatri mantra addressed to Savitri – the invisible sun of the night.

Here in Kashi, time is not determined by the clock, but by the cosmic cycle. As the sphere of Earth continues to spin, and at the same time circles her star, the rhythm of the city follows the movements in the sky – sunrise to sunset, the waxing and waning of the moon, twelve months of a year, and the six seasons.

Before the arrival of the monsoon, the stone steps are full of activities of faith and culture. But when the rains arrive, the waters rise and submerge the busy steps and some temples where water buffaloes swim in the inner sanctums and fish nibble at the idols carved in stone.

But again the waters recede, leaving behind deep marks on the stone walls, and return the submerged back to the faithful.

The poet wanted to take a boat ride every dawn. But only twice did he manage to hear the alarm of his wrist watch, and got up only once for the pre-morning sail. (He was always woken up by the spirited symphony of conches, drums and bells, but when it ended, he promptly went back to sleep.) So he took his boat rides in the evenings when the weather was more cool and pleasant.

And he always managed to find the same old boatman and the same old boat.

The poet enjoyed his daily boat rides. There were different rates for different distances. The length of a ride was measured by the distance between the series of stone steps which led to the river. All the steps had names. Rectangular nameplates painted on the stone walls in colours of gold and black sparkled in the sun like a fresh coat of glazing varnish.

After a few days, whenever the poet climbed down the steps closest to the guesthouse where he was staying, and arrived at the swaying depot of boats, the other boatmen didn't try to lure him, rather they helped him to spot *his* boat. Sometimes he had to wait till a heavy load of pilgrims and tourists disembarked from *his* boat.

The poet didn't want to share the boat with anyone and never bargained for a lower fare. That attitude pleased the boatman. The sense of loyalty that he reserved for *his* boat was also noticed.

The poet chatted with the old boatman who was a thin muscular man with deep, sparkling, marijuana-eyes.

The young boatman often adopted the role of an unwanted guide and offered unsolicited information – what movie was shot where – which didn't interest the poet. He felt more engaged by the words of the old boatman, who knew many stories and anecdotes about Kashi, and who described the city as being 'older than the universe'.

The old man had a poetic perspective, and hadn't grown old in vain. He had gathered wisdom, gravity and a mystical consciousness. The poet listened to his remarks with great interest, and even joy.

Most notable of the remarks was when he had said, 'Most of the street dogs you see in Kashi carry an expression of repentance on their kind faces. They were all humans in their previous births – rich, powerful, religious but corrupt. They misused their influence to exploit and harm others, and wasted the opportunities to do good. But the laws of karma don't accept bribes, so now they have a dog's life, and will do so for another hundred lifetimes. But sometimes, at the right conjunction of planets, they remember who they were, and what terrible wrongs they did. You can see that in their sorrowful eyes, as they deeply repent with a vacant stare.'

9

The next evening the old boatman asked the poet whether he knew the story of King Vikramaditya, the priest, and the lame black dog.

The poet said he didn't. The boatman remarked he had thought so, and explained that the story has been passed on orally through generations. He had heard the story from a great sage who used to come to Kashi when the boatman was young.

'Do tell me the story,' the poet requested, and the boatman said that he too wanted to pass on the story to the poet as a gift because it is said that the gift of wisdom is the greatest gift.

Then he began to narrate:

'King Vikramaditya, known for his sense of justice, allowed common subjects to come to his court and speak of their grievances.

'One day a lame black dog arrived and said that he was sleeping peacefully curled up near a temple when the priest of the temple trampled on his leg and then abused him as he kicked him away.

'The priest of the temple was summoned and he explained that the diseased dog had suddenly appeared in his way as he was returning from the river and he had accidentally touched him. Just because of that the priest would have to take another bath before he could begin the morning rituals, so he had lost his temper.

'King Vikramaditya said that the priest had committed an offence and asked the lame black dog to pronounce the punishment.

'The dog thought for a while and said that the King should promote the priest, and make him the head priest of the entire kingdom!

'The entire court started laughing at the dog. A minister remarked that the dog was mad. The priest couldn't contain his joy.

'King Vikramaditya pondered for a while and said that the dog's decree of punishment will be fulfilled and appointed the priest as the head priest for a year.

'When the drama got over, King Vikramaditya asked the dog to join him privately at his lotus garden.

'"I have this feeling that there is wisdom behind what you asked for," the King remarked to the dog.

'"There is no lofty wisdom, my King," replied the dog, "only vengeance."

'Then the dog explained that the crime that the minor priest from a small temple had committed wasn't such a great offence. But when he would become the head priest of the entire kingdom, his sphere of responsibility would be immense. Since the priest lacked the merit to discharge such responsibility fairly, he would make many mistakes and become arrogant, corrupt and selfish. Then the karmic laws would punish him suitably, and he would be born as a dog for a thousand years, and only then he would know what a dog's life was, and only then would he serve his years in repentance.

"That's a unique way of punishing someone who has harmed you," said the King, "but how did you know all this?"

"My King," replied the lame black dog, "I know this because a hundred years ago, I used to be the head priest of this very kingdom."'

10

The story delighted the poet and he couldn't stop himself from laughing.

'Why are you laughing?' the old boatman asked almost angrily. 'Didn't you see the wisdom hidden in this story?'

The poet said it was a humorous story. The old boatman replied very seriously, 'But I want you to understand the truth. So allow me to tell you another story so that you get it –

'A King and a beggar of the same kingdom died at the same moment. They found themselves standing in a queue. Chitragupta was judging the karma of the deceased and was pronouncing his judgment.

'There was no special treatment for a King, and he was standing right behind the beggar in a long queue.

'Then the beggar's turn came. Chitragupta looked at the *gupta chitra* – the secret picture that mortals draw in their souls through their actions.

'"One thousand lies, you have spoken," Chitragupta told the beggar, and then added, "You will go to hell for a single day, and then return to Earth as a farmer."

'Then the King's turn came. After looking at the King's soul, Chitragupta said, "You have spoken one lie. You will go to hell for a thousand years, and return to Earth as a beggar."

'The King was aghast with the judgement and protested. "How unfair is your judgement! The beggar before me had spoken a thousand lies and he got a single day in hell. And I with just a single lie get a thousand years in hell. How grave is this injustice!"

'Chitragupta remained unmoved and said, "Yes, the beggar had spoken a thousand lies, and you had spoken only one. But you were the King with your circle of privilege. Your sphere of responsibility was far greater."'

11

The poet understood the teaching behind the symbolic stories that the old boatman wished to impart to him.

'Yes, I get it,' the poet told the boatman. 'As the circle and sphere of your life becomes more privileged, you need to be more fair and responsible towards your duties.'

The old boatman gave a hint of a smile and said, 'Yes, that is it. Remember this, and also remember, you have nothing to fear except your own actions.'

12

The boat headed towards the crowded bank. The boatman shouted at his younger companion – at the other end of the boat – when he allowed the boat to drift too close to the bathers. The local children, who played a dangerous game, worried him. They would swim purposely near the hard-hitting oars and would dive at a precise moment to avoid contact with the wood. Then they would surface behind the flowing boat with a surge of triumph and laughter.

The boat approached the feet of the steps, muscled its way through the confusion of wood, and hit against the stone.

The poet balanced his stance to counter the wobble. He took out his wallet and paid the usual amount. But for some reason the old boatman wanted to return some change. 'You never bargain. It hurts my conscience.'

The poet smiled and told him to keep it as an advance for the next ride. Then he gestured a goodbye, thanked him for the stories and leapt onto a step.

13

The poet had discovered Café Mona Lisa within the network of lanes and bylanes of Kashi. It was a small little café that could barely fit in a few tables. The walls were decorated with reproductions of the famous painting in three predominant colours: green, blue and red. Dim cane lampshades surveyed over each table. The large menu offered a worldwide variety of food with an inherent Indian flavour.

The poet always ate at the café; the food didn't make him sick. Usually he managed to get a table to himself but sometimes he had to share his table with old hippies and tourists, who chatted with him about Benares and the world. Sometimes five different languages were spoken simultaneously against a background of trance music and temple bells.

The poet reached the café and found an empty table. The owner of the café recognized the loyal customer and came up for a chat. He asked those deeply personal questions – what do you do in life, are you married, how much does it take to live in a city – which the poet tried not to answer.

After he had eaten, he ordered tea, lit a cigarette and made plans for the next day.

Then the poet took out a notebook from his daypack and started filling it with random thoughts: 'I am a solitary traveller easing into myself within the sphere of this old city that enchants me with elements that don't exist within the realities of my urban life. But I am also aware, that beyond the visible and the invisible charm of the old city there are numerous problems stemming out of corruption, cronyism, sycophancy, nepotism, callous authorities and poor infrastructure. And there is even a betting syndicate run by goons who give odds upon the number of corpses which were likely to be burnt in the main riverside cremation ground of Kashi.

'The smile of irony strikes me. But the real essence of Kashi is beyond the domain of the good and the bad, and closer to the sense of timelessness and eternity.

'In *Maitri Upanishad* there is a poetic advice: cross to the farther side of the space within the heart.

'Kashi is the sphere where one can make that crossing.'

14

Next morning, loud knocks on the door startled the poet. It was the young lad from the nearby tea-stall. 'I came at nine but you were still sleeping,' the lad said and headed for the desk.

The poet had arranged for the breakfast and the morning newspaper to be delivered to his room. The tea arrived in an aluminum kettle along with two arch-shaped buttered toast sprinkled with pepper and sugar, a boiled egg and a banana.

The poet looked at his watch; it was well past ten. He had been reading and writing till dawn. He had only slept for four hours.

'The tea is hot but the food has gone cold. But it's not my fault,' the lad said and poured tea into a plastic glass. Then he took out a paper plate from a bunch that the poet had bought, laid the breakfast, crumpled the newspaper wrapping and put it inside his pocket.

The lad saw an empty water bottle and asked the poet whether it had to be filled. The poet said no, but the lad insisted, 'Don't worry, the water is clean and good.'

Then he explained that their tea shop also used the same tube well water that was used to fill the packaged bottled water that the poet had bought from the nearby medical store.

'My friend seals the bottles with a hot wire. He gets twenty rupees for every forty bottles. Why pay ten rupees for it? I will get you the same water for free.'

The poet thought for a moment and handed him the bottle.

The lad returned after a few minutes. He noticed the slim train-guide on the table and asked the poet whether he was leaving.

The poet told him that he would be going to a shop nearby to get an e-ticket and he would leave the next evening.

'Will you ever come back?' the lad asked.

'I think so, someday, after a few years.'

'By then I will have my own shop,' the lad said with confidence. 'But I won't toast all the stale bread and sell them to my customers.'

The attitude of the lad pleased the poet. Progress is made by the spirit that wants to do something better. He wanted to reward the young lad. He went to his backpack and pulled out three cotton T-shirts which he had bought. 'Choose one,' he said.

15

It was late in the afternoon when the poet returned to his room with a reserved train ticket.

He had visited the main temple and been to Sarnath – where Sakyamuni Buddha had delivered his first sermon to five listeners.

It was a hot day. The poet was drenched in perspiration. He hadn't shaved for three days; the itchy sensation on his cheeks created by the humid condition irritated him.

The water from the shower didn't burst forth, but died into a trickle. There were no buckets to trap the water. The poet feared the water would dry up any moment – he hastened to clean himself.

While shaving, the poet noticed that his finger nails needed to be cut; he had a nail-cutter in his utility kit. He cut the nails in his room, gathered the fragments and dropped them into the plastic wastepaper basket that he had added to the room.

Then he put on fresh clothes and went to the balcony. It was nearing evening. The river was getting crowded with boats.

Now it's time, the poet thought, to take the final boat ride of the trip.

16

This time round the poet asked the old boatman to take the boat further north. He wanted to linger in front of the burning ghat steps – the riverside cremation ground that exhumed smoke and ashes, which rose in the air, drifted over the river, mingled and dissolved in the water.

A corpse was being cremated. It was wrapped in white cloth. The rituals were being performed. Then a man lit the pyre.

The poet was thinking of death when the old boatman told him to visit the temple of Kedara (which literally means 'field') – the oldest temple in the city. The riverside temple had survived the cycle of construction and destruction initiated by the kings of history; its absence in the guide books and a plain entrance had kept it away from the crowd of pilgrims.

The boatman dropped the poet at the steps named after the temple. He bid a final goodbye to the boatman, climbed the steps, took off his shoes and entered the shrine.

The air felt more ancient. Fiery deities carved on black stone walls spoke of a time when gods were fiercer. The primal light of oil lamps lit up the darkness. The floor was cold, slippery and wet.

The poet approached the inner sanctum. The tempo of drums, the announcement of conches, the fierce tolling of bells and chants in a deep voice permeated the atmosphere.

Moved by the experience, the poet returned to his room in a mood of reflection. He was still thinking of death. His memory continued to play back the images – the corpse, the wood, the burn, the fire.

17

It was night. The poet was walking towards the ruins. From a distance he could see the fragmented structure of stone, glass and steel. It was an oddly shaped geometrical structure with curved corners, spires and domes.

He entered the ruins. There were many broken walls within an outgrowth of roots, vegetation and shadows. Then he sensed he was lost. A feeling of anxiety gripped him.

A group of hoodlums accosted him out of the darkness and asked him for something he didn't have. There was a series of unclear images before the skirmish. The hoodlums pressed him down on the ground. He felt a deep stab.

A curious sensation of blood gushing out from his chest caught him by surprise. And at that precise moment he knew, he was about to die.

18

There lies a twilight period between dreams and awakening. After a dream is over, the images slowly seep into conscious thought and then into memory.

When the poet woke up from the dream, he felt a wave of buoyant lightness engulfing him.

The very moment in the dream when he felt the dagger wound, he had realized that he was about to die.

But there was no fear nor pain, but a pleasant sensation of coolness and of flow. Every burden, weight and pressure seemed to evaporate, as if there was no longer any root or gravity. He was filled with a strange emptiness of pure joy and lightness.

The after-effects of his dream did not abate. It was a beautiful experience, of dying, of being emptied of everything.

For an hour, he could sense the lightness. His mind went blank. Then the awareness of his self began to change. Thoughts began to return.

19

There is a sublime difference between knowledge and wisdom. Both come from experience and understanding. But *knowing* and *realizing* are not one and the same.

There lies a childlike joy in knowing – it changes what one knew and thought of, but wisdom has the effect of changing the way one is aware of one's self. It transforms awareness of thoughts and of feelings.

The effect of a realization doesn't go away, but remains forever.

That's the root of wisdom – realizations which change the self, and modify response for the better.

The poet now thought: men live as if they are immortals but the wisdom of death frees a man. He learns to discriminate between the trivial and the real, between the surplus and the essential and begins to understand the true value of life.

Deep down death signifies a transformation, a metamorphosis. Death is not the opposite of life; life is birth, living and death.

20

Next evening the train was crossing the bridge. The poet was near the window, looking at the distance.

After a while, the city, the river and the bridge receded and became a gaze of his memory.

The poet had bought a bamboo flute from a child vendor at the railway station. He didn't know how to play the flute. He fiddled with it and tried to play a tune.

There was a little girl with her family in the compartment. She was watching the poet's effort with amusement – the odd noises of breaths and sounds, without rhyme or melody, made her smile.

The poet noticed the happy face of the girl; he pressed on with his lungs and shifted his fingers like a seasoned flautist. The more he failed, the happier she became. Soon everyone in the compartment was laughing.

The train sped through the brown dusty expanse, sprinkled with green fields of agriculture, a procession of steel towers – with their arc of power cables – and a slow moving horizon.

The poet looked out of the window, and felt the warm breeze against his face. He savoured the moment and closed his eyes.

From the sphere of the old city, he was now returning to the sphere of his home – the metropolis.

THREE

Centre and Periphery

1

Three years later, on a Sunday afternoon, the poet was staring at the horizon. The windows of the high-rise apartment looked southward. The metropolis was humbled and defeated by the rain. From the seventh floor everything looked deceptively simple and tranquil. The rain became heavier. Distance became hazier. The faraway towers loomed as silent phantoms. Down below, the veins of the living metropolis, appeared withdrawn and subdued.

The poet had watched the rain arrive from a distance.

When the rain had hit, the protrusion of the window air conditioner was assaulted by the shower. The sharp metallic pounding against the aluminium soon eased into a consistent rhythm. A cool dampness touched the poet's face. Something old and familiar swelled and blossomed, an awareness or a frame of mind that soaked and nourished him with a deep love for distance.

2

The poet hadn't noticed when the rain had stopped. He was at his desk, scrutinizing some papers, when his mobile phone rang. It was his grandfather. He told the poet about the day of his arrival. (His grandfather visited him twice a year.) The poet looked at the desk calendar to confirm his suspicion. A moment later, he told his grandfather to collect the keys of the flat from the neighbour. 'It's a weekday. I might not make it to the airport.'

The poet's grandfather had now become the editor of an astrological magazine. He had also written a couple of books. They had been well received within his circle.

His grandfather left after a few days. He had to attend an astrological convention in another city. But he didn't speak only of astrology. He compared the old times with the new and wore his sweater even when it wasn't that cold. It was a sign of advancing age.

But the poet's grandfather didn't think like an old man – he never used the phrase, 'during my time'. He always liked to say, 'one's time is not over until one is dead'.

3

The poet went to the airport to see his grandfather off. While driving back he could imagine his young grandfather walking amidst the strange honking of vintage cars. He was studying to be a doctor. A bright student, he belonged to a large family of four brothers and two sisters. There was also a day when a photograph had been taken on the roof of a building. The poet's great-grandfather was sitting in the middle. To his right was the poet's great-grandmother, to his left the two daughters, while the four sons stood behind them. The youngest was the poet's grandfather. The faces in the black and white photograph had been shaded with a tinge of reddish brown – the colour with which time polishes photos. They were all strangers to the poet.

The grandfather used to teach a girl of seventeen. She had many pets – three dogs, a parakeet, a few lovebirds and many rabbits. She was also entertained by the crazy antics of her aunts who spiked the evening milk of the talkative old widows in the family with opium and threatened to devour diamonds to attempt suicide. (There was a wrongly held belief that whoever swallowed diamonds was sure to die.) The master and the student fell in love. Both families didn't approve of the marriage due to differences of caste. But the lovers were determined. They abandoned their families and got married. They were shunned by their families. But a year later, the couple was welcomed back into the joint family upon the insistence of the poet's great-great-grandmother. She had fallen ill and the family had to respect her death-wish. When she died, the wives of grandfather's brothers literally fought amongst themselves to snatch her diamond earrings. When the poet's grandfather had told the poet of the incident as an example of horror, his tone and expression had brought about the gravity of the evil. Due to the snatching, the pillow was stained by drops of blood. The dead great-grandmother had to have her ears bandaged before she was carried to the crematorium.

Like it happens, the death of the elderly person created many differences among the brothers. Petty squabbles and deep conspiracies to seize properties and the keys to the safe dominated the atmosphere. The poet's grandparents were denied of their rights and forced to leave their ancestral home. The poet had heard of their struggles and tenacity. He imagined the simple little house where his mother had been born. There were two children. The poet's aunt didn't cross her teens. She always smiled in the photographs. She had died of a rare disease whose name was too long for the poet's mother to remember. After many years, the poet's mother returned from abroad without having completed her course due to homesickness. Then she fell in love with a young man. The man's parents had died when he was a child. He had been raised by his bachelor uncle from whom he had inherited his wealth. The young man's handsome looks, stories of travel to places of religious importance, a blue convertible and a mildness in speech had won him many admirers. But he fell in love with the poet's mother. They were married against the wish of the father of the bride – the poet's grandfather – who had begun to doubt the character of his would-be son-in-law after studying his horoscope. But his daughter was adamant. The marriage had to happen. The couple seemed to have found happiness. The wedding photographs were full of smiling faces.

4

The poet was now twenty-four. Like everyone else, sometimes he liked his work and sometimes he didn't. But he knew the sphere of print journalism was ideal for him – he could stay faithful to the printed word, keep alive his inquiries and write occasional poetry. He also aimed to send his poems to a publisher someday and had started collecting them in the form of a manuscript.

The poet was one of those people who got better at work when given more responsibility. His boss, a vociferous lady with a kind heart, had identified this trait in the poet. She gave him added responsibility that helped him to evolve: he understood there was no greater high than to do one's work well.

The poet stayed neutral amidst the undercurrents of ego battles in his office. He knew his place and behaved in a friendly manner. By doing so he was generally well-liked. Everyone got to know him as calm, sensible and dependable.

5

Every Friday evening, the poet played pool with a fellow journalist who was training to be a political analyst.

The poet was staring at the green table to figure out the best option. Just before playing his shot, in reference to an ongoing conversation, he remarked, 'Time is such nowadays, it is forcing everyone to think.'

The political analyst had smiled. 'Yes, people have to think when things are not going that great.'

'True,' the poet agreed, 'but it is more than that,' he added.

Later at a restaurant, while drinking tea and smoking cigarettes with his friend, the poet elaborated on his statement. 'I feel a great shift is taking place. Old games are no longer working. The general level of intelligence is increasing. I feel there is a growing desire for honesty, for fairness, for compassion, for transparency and for truth.'

'I was reading an article the other day,' the political analyst said, 'it predicted, the twenty-first century will belong to the right brain. The dominance of the left brain is on the decline. Analytical thinking is now secondary to ideas, creativity and conscience.'

'The signs are everywhere,' the poet said. Then he leaned back to reflect.

It was an idea of the poet that human consciousness evolves with every new generation.

What had so long been accepted as common occurrences, were no longer accepted. What had been looked at as a family affair (domestic violence, forced marriage etc.) was being looked at as an act of criminality.

The mindset changes with every new generation, trends emerge, personal and social behaviours get modified.

But more importantly the change in society takes place with the change that happens within ourselves.

'You understand what I mean?' the poet asked.

'Yeah, I do. We are sick and tired of selfish egoistic mindsets. They appear unwise, arrogant, narrow, corrupt and outdated.'

Then the political analyst pointed out that one of his friends had done a survey on social media posts. And he had found that there had been an increase of posts which demanded justice, evolved behaviour and change; it seemed as if people were collectively 'waking up' and venting their frustrations through Facebook, Twitter and YouTube.

'This is not limited to the virtual world,' the political analyst added. 'There has been an increase in apolitical street protests as well.'

'That's happening because of the mind shift,' the poet remarked.

'Do you think it's happening all over India?'

'It's largely an urban phenomenon, but it's more widespread,' the poet replied. 'It's happening all over the world. Different issues in different countries. A paradigm shift has begun.'

6

For the past few years, the poet had protected his solitude, using it as a bait to trap truths. He read many books, thought over many things, tried to remain calm and balanced, tried to be fair to everyone and made pages and pages of notes. He was making progress, but nothing yet had come to a final conclusion with regard to his inquiry into the human self. But he knew the right idea would occur only when one becomes worthy of it. He kept his faith in time and uncertainties, and waited patiently for his own moment of truth.

After work, except for Fridays, he kept to himself and refused to turn his apartment into a den of revelry. He met people outside – at cafés or pubs – and kept to his policy of 'not too near and not too far'.

Only his close friends from his college years were always welcome, irrespective of work pressure, moods and late hours of the night.

From time to time, the poet's friends visited him. Their presence gave birth to the happiness that only a friendship free of ego can offer.

But his friends had noticed that the poet appeared more withdrawn than before, and one of them had asked why the poet hardly appeared on Facebook. The poet had smiled and said, 'I just feel like staying within myself nowadays.'

The poet had been in a state of mind that was somewhat detached from the world around him. He was going through a phase when even the slightest attention appeared somewhat intrusive.

The prospect of adventure also failed to excite him. He had ample opportunities, but somehow he didn't want to disrupt the feel that he had created within himself. He also chose to keep a distance from the couple of young women who had moved into his sphere, eager for love, making slow phone calls, brushing their hair with their hands, cracking jokes whose real purpose was to put forward their best smile.

He avoided everything that was surplus – boring social functions and parties thrown by his mother's friends – and kept himself rooted to the real, the essential and the vital.

7

It is important not to have an enemy at home. No one knew it better than the poet – he had suffered enough domestic disturbances in his life to have understood the value of peace. But now when he returned home, he felt he had returned to his centre. There was only a peaceful sense, spaciously enhanced with music, meditation and the fragrance of incense.

The poet had added three new habits to his life – meditation, green tea and traditional Tibetan incense.

Every morning he would do a series of stretching exercises and fine tune his rhythm of breathing. And every night, before going to bed, he would meditate.

He practised the *Mahamudra* or The Great Symbol, read the ancient spiritual texts and kept a record of his experiences. His text of reference was *Mahamudra: The Moonlight – Quintessence of Mind and Meditation*, translated by Lobsang P. Lhalungpa.

The poet had mastered the *Mahamudra* posture. He sat crossed-legged in the lotus posture, straightened the body and the spinal column, placed his hands below the navel, bent his neck, pressed his chin against the Adam's apple, placed his tongue against the roof of his mouth, focused his gaze at a distance of five and a half feet, took normal slow breaths and followed the six rules of Tilopa, the ancient Indian master:

> 'Do not imagine, do not think, do not analyze,
> Do not meditate, do not reflect;
> Keep the mind in its natural state.'

The purpose of actual meditation, as the poet gradually understood, was to stabilize the attention of the inner awareness devoid of forms and thoughts, and to avoid clinging to cognition or mental formulations.

The poet also wrote down a quote from *Boddhicittavivarana* on a yellow sticky note and stuck it on his bathroom mirror with cello-tape.

> 'The mind, separated from mental images,
> Possesses the innate characteristics of space,
> And this space-like meditation,
> Is regarded as meditation on emptiness.'

The poet had progressed well during the three years of regular practice. From a mere five minutes, when he had started, he could now meditate for thirty-five minutes. His inner awareness had grown and he felt more connected with his body. He could feel and think with greater depth. The grip of ego and negative emotions on his inner self had decreased. He had gained more control over his response. He felt agile, energetic and his self was infused with a sense of flow and lightness. His intuitive judgment of other people – their feelings, attitude and nature – had become better. He also recalled all his vivid dreams with striking clarity, when he woke up in the morning.

He had bought hand-rolled, maroon-coloured Tibetan incense and burned them at all strategic corners, so that the aroma would spread evenly all around the apartment. He no longer could suffer the toxic smell of cheap factory incense.

The sublime smell of the incense worked as a therapeutic agent. He thought, meditated and slept much better. And he always lighted them, not with a sizzling match stick, but his blue cigarette lighter.

The poet had also learnt to cook – he could make himself a great cup of tea and a good breakfast. He had discovered the most convenient ingredients for an all-purpose brunch-like meal – bread, butter, cheese, chicken sausages, sautéed onions, fried eggs, tomatoes, orange juice, apple or banana. He could pick and choose four or five items and manage an occasional lunch or a dinner.

He also tried his hand at real cooking with spices and vegetables by taking clues from his mother's recipe books.

But the results were disastrous. Everything had to be flushed away in the toilet, every time.

The horrific experience of real cooking – fresh and homemade – made him realize that life cannot be led without the help of other people. He still had to rely on the part-time cook – the homely lady – who prepared for him a proper meal.

The lady was a good cook but she suffered from a nervous disorder – she couldn't remain silent, she just had to talk.

She delivered a live commentary as she went about her work. 'The tomatoes are cut, now I have to cut the onions. The onions are not that fresh. Their quality has gone down. But their price continues to increase. I cannot understand all this. They cannot grow onions or what? My nephew said, all the good quality vegetables are sent abroad, we are left with the worst. Good heavens, what is going on? Even the mustard oil has lost its pungent fizz...'

She also had an opinion about everything. She kept asking the poet to get married to a nice girl who would cook for him. The poet had pointed out that the arrival of such a girl might also mean that she would lose her morning job. She had gone silent for a moment and had then said that it would be fine, she could always find another job, but one must have a companion. Seeing her stance, the poet had told her, most urban girls cannot cook, they have other things to do, and the fact of the matter was that he was very comfortable without a companion.

But the lady failed to understand, and continued to pester. Then one day, the poet's patience ran out. He told her angrily not to talk about anything else other than 'what is to be cooked' and 'what has to be bought from the market'. The poet had thought – his firmness would offer him respite.

But something else happened.

She took a new strategy and said, 'Don't worry, I won't talk about *that* anymore.'

8

Flashing ahead between successive Mondays, the poet's life fell into a comfortable relationship between the centre and the periphery – he had allowed himself enough space and time to do all the things which stirred his soul (poetry and inquiries) and his profession as a journalist also allowed him the opportunity to utilize most of the traits which were close to him (thinking and writing).

This kept him balanced. There was no conflict and contradiction between what he believed and what he did.

9

In his study, the poet practised writing articles by assigning himself some topic. Even if none were meant to be published, he knew he had to develop his voice and master the *art of thinking*, to think with words, than in response to images and to memory.

For example, if someone asks, what kind of work one does in life, then the answer has to be based upon memory. The images will arrive in the mind, and one has to recall them as they occur.

But, if someone asks, what has life taught you, then there is a pause, one has to work harder and think.

Not only think, one has to reflect, feel and sense.

The difference between the two – thinking with words and thinking with images – is also the difference between insights and opinions.

Something happens, one can make a comment, give an opinion, a point of view.

But when one begins to think with words, one has to raise questions – how, what, why – then one begins to go deeper and wider, insights take place, ideas occur.

The poet kept writing articles – it was a way to practise thinking, to organize thoughts and to receive ideas.

He had understood good journalism was not all about news, but about insights. The idealism of good journalism stems from its intention to go deeper, from the deceptive periphery, into the truth of the centre.

10

From time to time, the sweetheart stirred the poet's core by her absence. On certain gloomy evenings, he recalled the sweetheart in the same manner he recalled his mother whenever he had fever.

The poet also dreamt of the sweetheart – harpooning dolphins on a white beach, kissing him in a hotel room whose walls were made of water. And the poet would wake up from sleep, choking with emotions.

The sweetheart still meant something deep – she continued to be a symbol of his dream world, a string to stir his self, a trigger to his emotions.

The poet realized old love never becomes old. They continue to simmer.

11

Remembrance of the past often induces relief (from the memories of pain), regrets (for what could have been) and nostalgia (as things had been).

When the poet was awake he could avoid thinking about the sweetheart, or prevent her thoughts from affecting him but in the reality of dreams, he was completely helpless.

A part of his self brought her back again and again.

Sometimes the emotions released by the dreams were so strong that the poet had to empty himself, free himself, and move them somewhere else.

For doing so, he had to rely on poetry.

Poetry was the way to transfer the central feelings to the periphery of a poem.

> This cold afternoon I feel you within me,
> in slow momentum,
> love, perhaps, is in transit.
> There is something mysterious about it,
> the distance between us – dreamlike, intangible,
> measureless – as if, it doesn't exist!
>
> I still feel your love touching my mouth.
> Your memories still shift within
> like sleeping children,
> dreaming of something deep and happy.
> In my dreams I don't know what your symbol
> seeks to convey, but I know,

>   this cold afternoon I feel you within me,
>   in slow momentum,
>   love, perhaps, is in transit.

12

After the poet had taken his final university examination and left the city, the sweetheart had little time to understand the absence of her former lover. She was busy preparing to leave for Germany and was excited about her new life – finally she had broken away from the grim realities of her home. Finally, she was free.

Earlier, she had hoped that the poet might change his mind. But the hope soon faded and disappeared. On the day of her departure, she had called up the poet. 'When are you planning to come back?' the poet had asked. 'Probably, never,' she had replied. There was no weight in her voice, she had sounded excited and happy.

Deep down she was drawn by the prospect of finding new love – someone more practical than the poet – who would understand the importance of love, marriage and family, and won't exhaust himself foolishly in vague solitary pursuits.

13

Three months into her new life, the initial wave of homesickness that the sweetheart had felt on her arrival had matured into an ailment. She missed her home, her dog and the sun. She never liked the cold; even during the cruelty of an Indian summer she needed warm water to take a bath. Anything cold made her shiver in terrible discomfort.

The cold became an issue for her; she began to feel discouraged to continue her course, and no longer wished to become a language teacher.

While talking to her mother over the phone, she kept complaining about the weather. It was true that the sweetheart's nose never seemed to dry up, but the cold weather wasn't the real issue – it was just an excuse – her real problem was the coldness of loneliness.

Some individuals long for freedom without being prepared for it. And when they find it, they often feel lost; unsure of what to do with so much space and so much solitude. The freedom turns into loneliness, without direction or an emotional anchor. And they tend to drift, making mistakes.

The sweetheart's condition was similar. She wanted to fill the vacant feeling in her heart; the prospect of a relationship attracted her.

But to fall in love and to find someone to escape from loneliness is not the same.

The sweetheart engaged herself in two successive affairs, which left her confused and dissatisfied. Both her lovers were intelligent and understanding – the sweetheart liked spending time with them. They could talk well and make her laugh. But when things became a little more intimate, something in her mind always compared them with the standard set by the poet. Every movement and touch of theirs only reminded her of her past. It was a terrible experience – in her mind she was with the poet while there was someone else in her arms.

She cited vague excuses, and ran away from her lovers.

Deeply disturbed and unhappy, she finally realized that she didn't miss a man, she missed the poet.

Her torment became magnified in a cold, foreign land. Then she gathered herself and wrote a very long email.

14

The email surprised the poet. There had been no news from the sweetheart for over seven months, not even a single phone call. Then all of a sudden, an email arrived with 'I only love you and no other' just before the digital signature.

The poet could make out from the words that the sweetheart had written the mail in haste, trying, on one hand, to suggest how beautiful the campus and the town was, and on the other hand, complaining of the cold and her sudden reluctance to become a language teacher.

For the most part, the mail was riddled with questions, 'whether you love me or not', 'we can give it another chance', 'we can live together at least' et cetera. And the poet didn't know what to make of it. A few months back, the sweetheart didn't want to come back ever, and now she wanted to know about 'our future'.

The mail put undue pressure on the poet – he was just beginning to understand the sphere of print journalism, arranging the notes of his inquiry and organizing himself in his solitude. His frame of mind was different.

Circumstances have to support crucial decisions. A right decision during wrong circumstances can lead to failures. The 'time' has to be right for many things to work, and the poet felt that the time wasn't right for him to engage himself with the challenges of a relationship.

He was in no mood to renew a commitment that had been terminated.

The poet, however, couldn't stop himself from worrying about the sweetheart. He didn't have the new mobile number of the sweetheart, so he searched out the number of her university residence.

The poet was informed by a fellow student that the sweetheart had dropped out of her course and had left for India.

15

Two days later the poet received a call from the sweetheart. She was back at home, with her dog and her parents.

Then she explained she had lost her desire to become a German language teacher; she no longer wanted to do it, so the only option was to come back.

This pleased the poet – the sweetheart had done the right thing, one must never feel scared to make changes in life, however hard or difficult they might be.

'It's alright. You don't have to feel guilty for wasting the scholarship,' the poet told her. 'You have done what you had to. Take some time off and figure out what you want to do in life. It is vital to do something that you like. Otherwise, you will be unhappy.'

The sweetheart smiled – how happy she felt talking to the poet. She agreed to everything that the poet said, like she always did.

Suddenly she realized what she had missed in others was the 'sense' of the poet, that always felt more deep, pleasant, understanding and positive.

Then she spoke of her father who had also shown unusual support for her action that had displeased many.

'That's great news,' the poet said, 'at least you won't have to fight another war at home.'

'Yeah, I know,' the sweetheart replied, her voice slowing down. 'I was reading one of your poems. *Just a little bit of sense, a depth, and an understanding, can change war to peace, anguish to hope.*'

16

Calls from the sweetheart came regularly for the next three months. The sweetheart had returned to her previous work at the gym as a yoga and Pilates instructor while she thought about what to do next.

Then one late evening the sweetheart asked the poet, 'Do you think about us? Our future?'

'I had been very clear with you. I need space to pursue a few important things. I need to be alone.'

'I can wait.'

'Don't wait. I don't know how long it will take. Get on with your life.'

After a short silence, the sweetheart suddenly asked whether the poet was in touch with the virgin.

'She wrote me a mail, one month back. She is just a friend, a *friend*,' the poet emphasized.

There was another interval of silence and then, 'Tell me frankly, are you seeing someone else?'

The poet laughed for a moment and then somehow he lost his patience – nothing angered him more than being doubted like a liar or a thief. 'When I say I need to be alone, I don't mean I have another lover,' the poet shouted. 'I am not trying to find an excuse to turn you down. Why can't you just believe me? I am not lying. I am sick of explaining things to you. Just do what you have to and let me do what I have to. It's better that you stop calling me. I just want some peace. Just leave me alone and get on with your own life, just get on.'

Sometimes in life, things happen very quickly, and finish abruptly, before they are grasped and understood.

Even in her wildest dreams, the sweetheart could never disobey the poet. Staying true to her sense of dignity, she never wrote another email, and did not call the poet ever again.

A few months later, the poet got to know from a friend that the sweetheart, after a brief love affair, had married a young man, unknown to the poet, and had moved away to another metropolis.

17

Two years had passed ever since. Except for the joker and a few of his hostel friends, all the other friends of the poet, including the virgin, had fallen out of his matrix. (She had completed a Masters in Business Administration and had got a job in a multinational company.) Everyone was busy with their own life and was doing what they had to. Everyone understood this fact. No one held any grudge or grievance for not keeping in touch with each other. But they knew, whenever they would meet again, even after a gap of several years, they would start off exactly where they had left.

But they did follow their social network updates, which only gave a partial or peripheral, and perhaps even a misleading portrayal of their actual lives.

In one of the emails that the poet wrote to the joker, there was this line: 'We meet, only to drift away.'

But the poet, overcome by the truth of transience, had forgotten to see the recurrences which take place.

Not only does the near become far, as Goethe pointed out, but sometimes the far, once again, becomes near.

18

The beautiful was a friend of the poet who had always stayed in the fringes. The poet had met her at a rock concert during the early days of his college life. She was at the concert with another girl; they were being harassed by a group of drunk rowdies. The poet had noticed their discomfort and had asked them to join his group. That had saved them from their ordeal, and the friendship between the poet and the beautiful had begun.

A month later, they had met at a New Year's party where the poet had to manage her when she got dizzy after a few glasses of wine.

The beautiful called the poet, 'my saviour' and the poet called her, 'my headache'.

The beautiful attended a different college and hung around with a different set of friends. But there was a chemistry of friendship between them despite the fact that the opportunities for social interactions were irregular and witnessed long gaps of several months.

Friendship among men is different from friendship among women. A beautiful woman somehow finds it more difficult to find a trustworthy woman friend. Even though the beautiful didn't have much trouble finding male friends, she could trust and respect only a handful. She had collected many bitter experiences; those whom she had considered friends had surprised her with red roses and invitations to candlelit dinners. All her male friends tried to elevate themselves to the status of a lover, revealing therefore, an ulterior motive behind the façade of friendship. Such behaviour distressed her, forcing her to trust her intuitive vibes – she realized that she had to perceive, not with her eyes, but with her mind. Being perpetually on her guard, she preferred to wear an aura of aloofness that others misread as snobbery. But with the poet it was a different matter. She didn't feel the need to be suspicious of the poet. A natural spontaneity developed between them.

Like most beautiful women, she didn't consider herself beautiful. It didn't matter to her that people were forced to notice her wherever she walked, her long limbs displayed a certain uprightness that the poet also spotted in her nature. She was candid, straightforward, good-hearted and somewhat innocent.

'I am a typical Sagittarius and you are a typical Leo,' she had once declared. However, it wasn't only the sun signs and the party hopping palmists who let her know about herself. Even though she bought more shoes than books, she loved to read, listened to music, went for walks and often stared blankly at the horizon in the distance.

19

After the poet had graduated and left for his home, they hadn't communicated much with each other. After three years, the beautiful had once again made an appearance in the poet's life. She was now a flight attendant. Her elder sister had been a flight attendant for an international airline. She had influenced the beautiful to pursue the profession. The elder sister was now married to a businessman whom she had met during a flight to Rome. She was no longer a flight attendant, but a homemaker who owned her own designer boutique.

But the beautiful didn't work for an international airline; she only did domestic flights. Her duty on the flights to the metropolis where the poet lived had grown more frequent. Every time she had a layover, she seldom thought of anything else other than to meet up with the poet.

'What kind of books do you read? I haven't ever heard about these authors,' she had exclaimed after examining the shelves which the poet referred to as his library. 'Don't bother about them. Tell me what's taking place in your life,' the poet had said while pouring the choicest orange pekoe Darjeeling tea that he reserved for special occasions.

If anyone meets someone for just three hours, every three weeks, then it is likely that both of them will always remain fond of each other.

But that wasn't the reason why the poet liked the beautiful. The poet considered the beautiful as one of his like-minded friends – honest, straight-talking and free from scheming tendencies.

The beautiful trusted the poet and didn't hesitate to narrate things which she found difficult to cope with.

It didn't take too long for the beautiful to tell her story.

She had been subjected to moral conditioning since her childhood. She was made to remember that her family expected her to lead a life bereft of all the acts which may cause shame and embarrassment to them. Constantly aware of being judged, she had led a safe life and had carefully chosen the people she allowed herself to be seen with. Deep down, she was eager to enhance her experiences in life, but found herself constrained with the need to fall in love.

Then she graduated, started working and moved to a new city where she shared a small flat with another working woman who had left her husband after he screamed out the name of the neighbour's wife while he made love to her.

'Men are shameless,' her flatmate had told her. But the beautiful couldn't say anything, she had had crushes, but never a boyfriend or a lover.

After a year, the initial enthusiasm of work had been replaced by fatigue and monotony. She had visited every corner of the country, virtually lived in five-star hotels, ate exotic food every day, shook a leg in the best night clubs, met new people and lived the life of a flight attendant that was idealized in the fairness cream advertisements.

And whenever the beautiful saw those ads, she felt like shouting from her window and telling the world about the reality of it – stories of drunk and mischievous passengers, the hectic work schedule, the stomach cramps, the sleepless eyes, the low wages and the ever growing supply line of new aspirants. Nothing was 'happening' about the job – it wasn't a dream – it was just a job to make a living, full stop.

The beautiful always loved to use the word 'happening'. She tended to divide the reality of her experience under two simple categories of 'happening' and 'stagnant'. For her 'happening' was good and 'stagnant' evil.

After a few more months, stagnation settled in her life like a group of unwanted relatives.

The beautiful didn't feel satisfied with the work of a flight attendant. Yet she valued the independence that her job offered. But her emotive life had come to a complete standstill.

On her free evenings, she got tired of going alone to the movies, and watching cosy couples leaning towards each other. Whenever she saw a kiss, she felt a tingling sensation around her own lips.

The beautiful wanted a man in her life. She wanted to be struck with the 'lightning of love' and suffer the crippling strokes of romance. Nothing of that sort happened when she fell in love with a young co-pilot. He was within her sphere, he was smart, handsome and many other girls were after him. The sense of competition sparked her. Within a week, the young co-pilot was her boyfriend.

The family of the beautiful lived in another city. There was no one to spy on her. The young co-pilot lived alone in a flat; she wanted to move in with him.

It was entirely her decision. The co-pilot had resisted the idea. When he agreed, he had pointed out that the live-in arrangement didn't include any long-term commitment.

The beautiful was smitten with love, she agreed without grasping the consequences.

After a few happy months, fears began to surface within her. The beautiful heard, 'I love you', but she wished to hear, 'I will love you forever'.

She began to fear the day when her lover would leave her and disappear without a trace from her life.

'What do you think I should do?' the beautiful asked the poet.

Usually a woman only wants to be listened to. She doesn't like being cut short and advised. But the beautiful wanted the poet to say something.

'Tell me, what should I do?' she asked once again.

The poet paused for a moment. Then he said, 'Talk to him clearly. Don't hide your fears. Speak frankly from your heart. Be strong and see what happens.'

'See what happens' reminded her of a future that she feared. She found the feeling of intimate security missing from her life. She let out a deep breath, curled up on the sofa and clasped a cushion with her arms.

'Aren't you happy with him?' the poet asked.

'Yes, I am,' the beautiful said imploringly, then she paused for a moment and added slowly, 'I am at the peak and also at the bottom simultaneously.'

The poet understood the statement: the beautiful liked her relationship but wasn't comfortable with her 'status' in the relationship. She was neither the wife nor the mistress. She had no power, no voice, no right. She didn't feel secure. The uncertainty of 'what will happen next' made her anxious and unhappy.

After a while, the poet said, 'As a friend, I can only ask you to challenge your fears and stay prepared to counter any situation. The rest, you will have to do it by yourself.'

20

During the subsequent visits, the beautiful avoided talking about her private life and the poet never pressed her to know what was happening.

When her mood was down, she craved for homemade food, and used the poet's kitchen to prepare a simple meal.

But when her mood was upbeat she was quick to announce her enthusiasm for going outdoors. The shopping streets, the cinemas and the restaurants attracted her with the constant flow of herd warmth.

'Let's go out for a movie,' the beautiful proposed.

The poet heard the name of the movie and flatly refused.

'Come on. It's just some entertainment,' the beautiful pressed on.

'I am tired of entertainment. It causes me distress.'

'Just for my sake, distress yourself. I really want to see that movie. And don't forget, watching movies is the only way I am going to learn how to direct one.'

That's what the beautiful wished to do. To direct a film. Earlier she had wanted to pursue theatre and had been involved in a few amateur plays during her college years. But later, after watching the quality of popular movies, she thought she could do a much better job. She had a degree in accountancy, worked as a flight attendant but her true calling was in the arts.

And apart from all this, she was the live-in partner of a co-pilot who said 'I love you' but never said 'I will love you forever'.

21

The beautiful had got sucked into everything before she could even realize what was really happening.

Yet she was getting accustomed to the patterned confusion of her life, without her own will, or consent.

At one level, she had her job, her need, her relationship; she had her fears, her stress, her worries. But at a deeper level, she felt she had been hidden away from her own life.

As if her centre, her soul, her spirit, had sunk away somewhere deep, and she failed to get it out into the life around her, the periphery.

One year ago, a weekly magazine was doing a feature article on young working women. She was also interviewed as a modern young woman with a mind of her own. She was busy, had her own life, took her own decisions.

But now, she had this strange feeling that all her decisions weren't really hers!

Not a single decision of hers was well thought out, the will of her heart had never been used!

She simply got carried away and responded to situations – to trends, to opportunities, without ever thinking whether there was any actual need to respond to them or not.

The beautiful didn't like what she had done with her life. Perhaps, it was finally time to make changes, and get herself engaged in her 'own' life!

But what would she do? She could leave her job, get away from her uncommitted lover and go back to school to learn the art of film direction, and follow her hidden dream.

The idea sparked her spirit, but somehow she couldn't find the courage to put her thought into action.

Her will vacillated with her ever-shifting feelings.

One day she would think of making a fresh start, but on another day, she would find herself lovingly making dinner for her lover, and feeling deeply fulfilled on witnessing his smile of contentment.

And she knew, with a whiff of unease, that her next step would be decided by her lover, rather than her own self, and only then would she take any step. And the poet was also somehow certain that the lover would agree to marry her eventually, and the beautiful would forget about film direction, be a homemaker and start a family.

In life, more often, one regrets not doing something, rather than regretting something one has done.

And years later, the beautiful may just develop a silent regret, and wish that she had taken a different decision, or rather, her lover had refused to marry her!

But again, she might not have that regret and feel fulfilled with the life of a homemaker.

But knowing the beautiful, the poet feared that her destiny was perhaps leaning more towards eventual regret, than otherwise.

22

During her next visit, the beautiful chose to carefully survey the poet's 'library' and picked out a slim book. It was *The Undiscovered Self* by Carl Jung.

She read a few pages and remarked, 'This is so difficult to read. How will anyone understand what he is saying!'

The poet placed the plate of chicken momos that he had heated in the microwave on the table and said, 'On the contrary, it is the most accessible of all his books.'

'What does it say?' the beautiful asked.

'Importance of self-enquiry and self-realization as the means to transform the psyche.'

The beautiful went silent and wandered into her thoughts. Then suddenly a thought sparked in her. In a tone of hopeful expectation, she asked, 'Does he explain what is the self?'

'He speaks about the existence of an obvious duality of the good and the evil within us. But nothing more than that. That's why the book is called *The Undiscovered Self.*'

The poet's words extinguished whatever little interest the beautiful had in the book. She slipped it back into its place on the shelf.

She turned around and asked the poet, 'Do you understand yourself?'

The poet waited to finish eating a momo and said, 'I do, in fact, to a certain extent.'

'I don't understand myself,' the beautiful said, 'I can't seem to find my centre. I forever seem to be standing at the periphery.'

After saying those words, a mood of reflection engulfed her. Then she emerged from herself.

'I only trouble you with my problems,' she said.

Before the poet could react, she cut him off. 'I don't want to be in a sombre mood. I don't want to be serious. Let's go out and watch a brainless movie. You will have to come, you have no option.'

The poet tried to suggest that he had added some new DVDs to his collection of world movies, and the beautiful could select any one of those, watch the movie and eat some Chinese food that he could order over the phone.

'No,' said the beautiful, 'I don't want to watch art movies. Not in the right mood. I want to go out, please.'

23

The poet dropped the beautiful at her hotel after the movie.

While driving back, the poet kept thinking about what she had earlier said. The two words, 'centre' and 'periphery', hovered in his mind.

He returned home, went to his study, read his notes and reflected:

> Within every individual there is a heart and a mind. (Both heart and mind are intelligent, having a 'reason' of their own.)
> Heart is sense, spirit and feelings – a spectrum of responses – from being selfish to being selfless, from sentimentality to soulfulness, from shallowness to depth.

Mind is propensities or traits – memory, intellect, thoughts and understanding.
Heart and Mind form the centre of a person.
And that centre comes out into the periphery (the life outside) in the form of human actions.
The constant play between the centre and the periphery is the very basis of life and of existence.

24

At the newspaper office, the next day, the poet's boss had an idea. She wanted to do a weekend feature on people who had decided to leave their jobs, give up their career midway and take the plunge to start something new.

She had pinned down four individuals who had made a successful transition from the familiar to the unknown, and she wanted to focus on their motivations and inspirations.

The poet was briefed about a certain person – a corporate economist, who had worked abroad for a multinational company, and who suddenly decided to return to India and educate street children.

After ten years of work, he had established three residential schools on the outskirts of the city.

He also helped village artisans, working in traditional forms of handicrafts, to sell their art through a website that he had created.

The poet had to interview the person. 'Concentrate on his mind,' the poet's boss had told him. 'I want to understand what prompts people to get back to zero, forfeit all that they had worked for and start all over again.'

25

The interview went off well. The gentleman, now fifty-five, told the poet that at the age of forty-four, he realized, time is not money, time is life.

'My wife died of cancer. It was a prolonged suffering. My own work seemed meaningless. I was down and depressed. But somehow, within all that, something came alive. You may call it the root of my soul.'

Like a star, a man is reborn through clouds, chaos and suffering. With the shift in one's feelings, the mind changes, the focus shifts. One begins to see things with new eyes.

'In hindsight, I have realized that a crisis is a tool of self-evolution. When crisis strikes you, it reveals your own poverty of understanding and your lack of ability to deal with it. A period of anguish and disillusionment sets in, that can make or break your life. But if you can utilize the crisis to search your soul and develop new ways of thinking, then that crisis gives birth to a new consciousness within you. And you can find answers to heal yourself for the better and do new things, which you had never thought of doing before.'

While listening to his words, a couple of lines of poetry formed in the poet's mind.

> In the season of darkness,
> the tree of self,
> grows new leaves
> and branches.

The poet wrote down the words in his notepad, and went back to the words of the gentleman.

'For the last ten years, through my work, I have also discovered myself. There is a sense within us that is truly ours. It is not ruled by self-interest; it doesn't respond to profit. It's a different aspect of the human self. It wants to give, than to take. When you can act out of that sense, then you create happiness. One has to do good to feel good. This is the truth.'

'But what about the prospect of starting all over again?' the poet asked. 'Didn't that scare you, all that hard work?'

'I knew it would be a long and tiresome battle. I was mentally prepared. I took it up as a challenge. And for doing so, I could understand that I had to respond from the better part of my own self and avoid the worst.'

'Work as a way of self-development,' the poet observed.

'Exactly,' the gentleman agreed, 'but for that one has to go beyond selfish motives, and start thinking of making a contribution. That's the attitude one must cultivate.'

26

Once the poet was alone in his apartment, he kept listening to the recorded conversation while sipping endless cups of green tea. Then he typed a transcript and printed a copy.

After all this, he felt like going out for a walk.

He picked up the keys, his pack of cigarettes, took the elevator, passed the usual watchman, opened the heavy gate with a noise, pushed the gate shut, more slowly, and ventured out into the street.

It was fifteen minutes past one. The air was cool. The streets were empty. The traffic lights were blinking yellow.

With the buzz of mechanical flies, a few fast cars approached and passed by – the linear sound, devoid of any highs or lows came near, peaked and then faded away.

The poet paused by a park to light a cigarette. Then he moved away from the mechanics of the main road and entered a low lit street.

The poet was still thinking about the person whom he had interviewed.

The person had shifted from being a corporate economist to being a social worker, from the profitable to the good, from *well-being* to something else ... *being-well*.

And *being-well* is the opposite of *well-being*, not in course with self-interest – ego-sense – well-being, but the other stream that doesn't want to *get* but to *do*, to give, to actualize, the other stream of soul-interest – soul-sense – being-well.

27

The idea sparked the poet – he had to write it down. He was out in the street; he had to get back home.

Without conscious intent, he had followed a pattern in his walk – a circular one. If he simply walked ahead, he would reach the high-rise.

The poet quickened his pace. A couple of street dogs cast attentive glances and walked towards him. A dog can sense a man's thoughts; the poet's presence didn't trigger suspicion; he was allowed to pass in silence.

Then a drama erupted. A howl reached the poet from the adjacent street. The street dogs behind him joined the chorus.

A trace of fear lingered in the air, a stalking anxiety.

At that very moment, with a chill passing through his spine, the poet had a sudden spark of insight. A sketch emerged in his mind's eye. He could visualize the pattern of the human self – its structure, its architecture, its design.

28

For the next eight months, the poet only attended his office, did the work he had to, and rushed back to his study to sharpen his understanding.

There was no one to disturb him at home. He could think through the evenings, till late into the night.

Except for phone calls, one or two casual visitors, the morning cook and the cleaning lady, nothing else fractured his time.

The beautiful had also vanished from his life. Her roster had changed; she had been assigned to a new sector and no longer flew to the poet's metropolis.

The communication between the poet and the beautiful had virtually died down. But she called a month later and gave the good news that her lover wanted to marry her, and the marriage would happen in a few months, and that she had decided to quit her tedious job and focus on her future.

'I always knew this would happen,' the poet said. 'I am sure you will be great wife and have a great family life. All the very best.'

Balance

1

The human self has come before all religions, nations and boundaries. But what is the self?

That was the fundamental question which had pursued the poet through the years and finally he had an answer.

The poet constructed a model, to depict the anatomy of the self – a skeletal invisible – that works through the human anatomy, and forms the very basis of humanity.

It took the poet three months more to organize his thoughts and study all the notes which he had made over the years. Then he decided to write a treatise titled *The Discovery of the Self and its Anatomy*.

After writing the treatise the poet didn't know what to do with it. Getting the treatise published in print was impossible. But thanks to the virtual world of the Internet, he could easily publish it through the *Youth India Culture Post*.

The online magazine – still funded by friends, had grown in stature over the years. Its story had also been covered by the mainstream press. The poet had done a lot of editorial work for the first two years, before he shifted to the role of a regular contributor. He had published eight essays and thirteen poems in less than four years.

When he sent an SMS to his senior and informed him about the essay, he was asked to submit the treatise immediately.

The treatise was published after two months.

2

The Discovery of the Self and its Anatomy

A Short History of the Self

The Rational Self

The fundamental behaviour of the rational, economic man is to pursue one's self-interest and well-being.

A rational person always seeks to maximize profit and consumption, in other words, ensure his *well-being*.

But in reality, not all human behaviour mirrors this kind of action. A man can act against self-interest. He can be selfless. But in the field of economics, this behaviour is termed as irrational.

The Irrational Self

Siddhartha Gautama was born in 563 BCE. He was the son of Shuddhodana, king of the Sakya's – one of the many royal kingdoms of ancient India.

According to *Buddhacarita*, the literary biography of Sakyamuni Buddha by the poet Ashvagosha, the Sakya queen Mahamaya gave birth to Siddhartha in the paradisiacal grove called Lumbini. Before conceiving Siddhartha, Queen Mahamaya had a dream in which a white elephant entered her body. The wise men of the kingdom had interpreted the omen as the indication of a son who would be a great leader, the most powerful of all kings, or a great teacher of truth. The word spread, and the seer named Asita came to the palace to see the royal baby. As the story goes, Asita had told the king about the spiritual destiny of his son.

Siddhartha was raised in Kapilavastu, the capital of the Sakya kingdom. He was well versed in the oral tradition and classical philosophy – the *Upanishads* are so ancient, that they were also *old* in his time.

King Shuddhodana didn't want the prince to become an ascetic. He schemed to entrap Siddhartha in a sphere of opulence and sensuous pleasures.

But his scheme failed. Siddhartha confronted human suffering in the streets of the kingdom. Death and misery affected him. He felt his awakening.

By that time, Siddhartha was also married and had a son. He was the prince of Sakya Kingdom but he wanted to seek a way to end misery. The sphere of his princely life conflicted with the sphere of his calling. Siddhartha had to take a decision. Either he had to remain in the palace, or venture out, to pursue his calling.

He also had another dilemma. What was greater – his responsibility towards his kingdom and his family, or his responsibility towards his calling?

At twenty-nine, he realized, the cause of his calling was greater. His sphere of life had to change. He had to leave the palace and begin his solitary quest to understand the causes of suffering and the way to end it.

One night, when the palace was asleep, Siddhartha took a final look at his sleeping son. Then he aroused Chandaka to saddle his trusted horse Kanthaka. Siddhartha and

Chandaka, rode out on their horses, and galloped through the night till they reached the periphery of a forest. It was around dawn when Siddhartha took off his jewels and dismissed Chandaka (who tried in vain to dissuade him) with a message for his father. In the message, Siddhartha spoke of the truth of transience, the ultimate parting in all human relationships, the uncertainty of how much life is at one's disposal, the necessity for immediate action and his unshakable resolve to fulfil his purpose.

From then onwards, Siddhartha, the prince of the Sakyas, began his life as an ascetic in the deep forest of Uruvela.

From the viewpoint of the behavioural assumption of a rational economic man, the human self can only pursue self-interest and well-being. Hence, the act of Siddhartha renouncing his royal well-being in favour of an ascetic life may be judged as grossly irrational.

The Natural Human Self
The Rational and the Irrational

The story of Siddhartha Gautama illustrates the *irrational self*, its existence within the natural self of a human being.

(A human being is what nature allows him/her to be. Nature isn't simply the sky, the wind and the mountains but also bone, blood and even an electron. A human being is as much a blossoming of nature, as is a flower, a tree or a tiger.)

So what is the natural design of the human self – the rational and the irrational? What is truth and reality?

Well-being and Non-well-being
Siddhartha as an Agent

Many economists have debunked the behavioural assumption of self-interest and well-being. The rational, economic man has been called the rational fool, but Amartya Sen has gone a bit further. Sen points out that a person's choice is not necessarily guided only by the pursuit of his or her well-being, a person as an *agent* may have important non-well-being goals and objectives. He makes a distinction between 'the agency aspect' and 'the well-being aspect' of a person. Sen writes (in *Inequality Reexamined; Chapter 4: Freedom, Agency and Well-Being*):

> 'A person's agency achievement refers to the realization of goals and values she has reasons to pursue, whether or not they are connected with her own well-being. A person as an agent need not be guided only by her own well-being, and agency achievements refers to the person's success in the pursuit of the totality of her considered goals and objectives.'

From this perspective, Siddhartha's choice of denouncing his royal well-being is no longer irrational. Before leaving, when he told Chandaka of his purpose and his resolve to discover the way to end misery, he displayed the 'agency aspect' of a person who has important non-well-being objectives.

In the 'rational sphere' of economics, the 'irrational' has now been rationalized.

It can be concluded that a self has a two-fold tendency that provokes him or her to pursue *well-being* and *non-well-being* goals and objectives.

Dostoevsky's Theory of the Self

The Underground Man of Dostoevsky/Siddhartha's 'Independent Wanting'

Maybe man does not love well-being only? – asks the underground man of Dostoevsky (in the novel *Notes from Underground* - a new translation by Richard Pevear and Larina Volokhonsky), who passionately differentiates between 'profitable wanting' and 'independent wanting'.

What Dostoevsky called 'profitable wanting' is another term to denote what is known as man's pursuits of well-being motivated by self-interest.

What he called 'independent wanting' is something that has to be understood.

The underground man of Dostoevsky points out, 'man, whoever he might be, has always and everywhere liked to act as he wants, and not at all as reason and profit dictate; and one can want even against one's own profit, and one sometimes even *positively must*'.

Dostoevsky argues that 'independent wanting' is more profitable than the profits one can acquire by following the reason of self-interest and well-being.

The underground man says, '...And in particular it (independent wanting) may be more profitable than all other profits even in the case when it is obviously harmful and contradicts the most sensible conclusions of our reason concerning profits – because in any event it preserves for us the chief and dearest thing, that is, our personality and our individuality.'

From Dostoevsky further insights can be gathered. Not only can a man defy self-interest and well-being for the pursuit of 'independent wanting', but this very 'independent wanting' of man is related to love and the dearest part of his self, that a man cannot violate, and if he does so, he violates himself.

He writes, 'Man needs *independent* wanting, whatever this independence may cost and wherever it may lead.'

From this perspective, Siddhartha defied 'profitable wanting' in favour of 'independent wanting'. He was aware of something that was more compulsive and dearer to him than anything else, the part of his self that craved for a pursuit that would fill him with a kind of love for which he was prepared to sacrifice his royal life and confront the unknown. Siddhartha knew the *profit* of this immaterial love that gave him strength and resolve, and he sought to fulfil himself.

*

Two Personalities of the Self

Elsewhere Dostoevsky advised: 'Try to be separate – try to determine where your personality ends and another's begins.'

The personality of 'I', craving for 'profitable wanting', and the personality of 'I', craving for 'independent wanting', aren't the same.

The 'I' of Siddhartha, when he says, 'I must strive for the highest good,' displays the personality of self that is different from the personality of 'I' who craves for 'profitable wanting'.

In other words, the two opposing goals – to pursue 'profitable wanting' and 'independent wanting' – create, modify and influence the attitudes which are required to accomplish them.

The Realm of the Spirit

Martin Buber: I-Thou and I-It

The attitude of man is twofold ... the *I* of man is also twofold – declares Buber in his poetic work titled *I and Thou*.

The essence of Buber's short work teaches that a twofoldness runs through the whole world, through man and through every human activity. *Thou* is the living centre of man that corresponds to the *Thou* of the Cosmos, the eternal living spirit. When a man forms a *I-Thou* relationship, or in other words, when a man confronts the eternal part of his self, his realizations provoke him toward actions which are interspersed with meaning, joy, creation, morality, aesthetics and happiness. If a man negates his relationship with himself, he negates the *I-Thou* relationship, then he is solely motivated with *I-It* (worldly pursuits), dominated by the will to profit and will to power.

> Man's relationship with the *Thou* is *I-Thou*.
> Man's relationship with the world is *I-It*.

From this perspective, Siddhartha had formed within himself an *I-Thou* relationship that mattered more to him than an *I-It* relationship. It also suggests that an action necessitated by *I-Thou* relation corresponds to the eternal Cosmic *Thou*, and therefore, is essentially spiritual in nature.

*

The Heart of the Self

Man is part earthly, part heavenly, and that a life of goodness enhances the heavenly part of man is an ancient Orphic belief that originated in ancient Greece. And by doing good, a man can increase the heavenly part of himself that the Chinese knew as *hun*, the soul of the spirit.

It was also during 5 century BCE that Taoism's original classic *Tao Te Ching* was written by Lao Tzu in some five thousand characters. In the *Shih chi* (Records of the Historian) written by Ssu-ma Ch'ien at the beginning of 1 century BCE, it is recorded that after meeting the old man, Confucius had said, 'Today I have seen Lao Tzu who is perhaps like a dragon'. During such times, there was a belief that man has two souls, *p'o* and *hun*. *P'o* is the earthly soul of the body and *hun* is the heavenly soul of the spirit.

Confucius differentiated between the petty man thinking only of profit and the good man of noble character who acts in accordance with one's genuine nature.

Jen, suggesting humanity, goodness of character, benevolence etc., is the foundation of Confucian ethics. The symbol that forms this character *Jen* means 'two human beings'. Confucius suggested *Jen* as the relationship between two human beings, the petty man (guided by the earthly soul of the body) and the good man (guided by the heavenly soul of the spirit). These two men, the petty and the good, live in every man. By practising acts of virtue and morality, controlled by *Li* (rules of propriety) it is possible for every man to increase the influence of the good man within himself, thereby lead a spiritual life and become a suitable seeker of the *tao* – the absolute void-like principle underlying the universe, combining within itself the opposing principles of yin and yang, signifying the path or the way, or code of behaviour, that is in harmony with the universe.

Within the Mahayana canon of Prajna-Paramita, one encounters that the essence of the Void dissipates within man as *pugdala ego* or the *selfish I* and as *dharma ego* or the *virtuous I*. In *Bhagwati Prajna-Paramita Hridaya*, the twin propensities within an individual 'I' are called the 'Two Egos'.

In the final part of the 'treasure teachings' or *terma* of the Vajrayana doctrines – orated by Padmasambhava to Dakini Yeshe Tsogyal, princess of Kharchen, in ninth-century Tibet – the way to manage the 'innate demon', and keep oneself aligned to the 'innate deity', at the time of after-death, is documented.

In more recent times, the Japanese monk Rosen Takashina of Soto sect in his essay *Controlling the Mind* (translated by T. Leggett; whose extract also appears in Conze's *Buddhist Scriptures*) writes about the human heart that has two aspects: 'They are the pure heart and the impure heart. But the heart in itself is not two; it is only classified in these two ways according to its workings.' The impure heart is the selfish egoistic heart, the source of suffering, while the pure heart is the natural heart which is characterized by wisdom, goodness and compassion.

The *Samkhya* system of Indian thought is based upon two kinds of forces or tendencies – *prakriti* and *purusa*. They flow through the self of man, as the lower and the higher self. (Arthur Schopenhauer, who was inspired by this idea, named the forces, *will of the species* and the *spirit of the species*).

In the *Zend Avesta*, two competing propensities existing within man are introduced in a doctrine of dualism.

The Hebrew term *Satan* (Ha-Satan) means 'obstruct, oppose' – an adversary to humanity (similar in idea to the Buddhist concept of Mara – a force in the human self that manifests as an obstruction to spiritual growth).

In Islamic theology, *Shaytan* means 'enemy/adversary' who whispers evil into the hearts of individuals, and is also connected to emotions which instigate a desire for selfish gains as against universal good.

In the Bahai faith, *Satan* is not an evil power existing outside humans, but signifies the lower nature of humans. Abdul Baha clarifies, 'Lower nature in humans is symbolized as *Satan* – the evil ego within us, not an evil personality outside of us.'

In the second part of *Serek Hajjahad*, The Rule of Community or The Manual of Instructions, one of the ancient scrolls discovered in 1947 by a young bedouin in a cave beside the Dead Sea, the 'two spirits in nature of man' is described.

In the *New Testament* (*Galantians 6*), a contrast is made between human nature and the Spirit that produces good qualities within man. *Galantians* declare, 'these two are enemies and oppose each other'.

In the *Bhagavad Gita*, characteristics of godly and devilish nature are expounded as two opposing potentialities which exist within the self of man.

Dhammapada says, 'Self is master of self', the *Bhagavad Gita* says the same (VI, 5), 'Self's friend is self indeed/So too is self self's enemy'.

*

Self Elementals

The Yoga of Non Ego, expounded in *The Tibetan Yoga and Secret Doctrines* translated by W.Y. Evans-Wentz, is one of the seven ancient texts that the book consists of.

The Yoga of Non Ego is also based upon – like all Tantras – visual imagery, mantras/sound and meditation.

The central theme of the yoga is to subjugate the lower self.

The lower self culminates as egoism that is fed by five elementals: Hatred/Anger, Pride, Lust/Greed, Jealousy and Ignorance.

The antidotes to the five elementals are imagined as five *dakinis* or aspects of one's consciousness carrying the five spears of Love, Compassion, Altruistic affection, Impartiality and Wisdom.

Or in another sense, *soul-consciousness* (love, compassion, altruistic affection, impartiality, wisdom) is aroused to dissolve egoism that creates *ego-consciousness* (anger/hatred, pride, lust/greed, jealousy and ignorance).

*

Ardhanarisvara – Half-man and Half-woman

In the *Samaya* school of Kundalini Yoga Tantra, the two opposing propensities are called *Aham consciousness* and *Idam consciousness*.

This is visually depicted as the symbol of *Ardhanarisvara* – a form of half-man (related to the sun – *pingala*) and half-woman (related to the moon – *ida*).

(*Susumna* is the central void running through the spinal cord wrapped around by *ida* and *pingala* channels)

From the point of experience, the right side of the body is man and the left side of the body is woman.

This form of opposing forces lies between every human being (whether male, female or hermaphrodite).

The human embryo grows and develops into two sides, left and right.

From the point of experience, the right side becomes the field of *Idam* or *soul consciousness*, and the left side becomes the field of *Aham* or *ego consciousness*.

Two Opposing Forces

A Short History of the Self points to a play between:

>*aham* and *idam*
>*will of the species* and *spirit of the species*
>*lower self* and *higher self*
>*innate demon* and *innate deity*
>*pugdala ego* and *dharma ego*
>*impure heart* and *pure heart*
>*petty man* and *good man*
>*p'o* and *hun*
>*soul of body* and *soul of spirit*
>*human nature* and *spirit nature*
>*earthly soul* and *heavenly soul*
>*I-It relation* and *I-Thou relation*

profitable wanting and *independent wanting*
selfish I and *virtuous I*
well-being and *non-well-being*

This dual tendency of the human self – intertwined like two coiled serpents or the double helix of DNA – are the two forces behind every human pursuit, the leitmotif of life and of existence.

The clash between them is the root of ethics and of moral sense.

Anatomy of the Self

Truths cannot be created or invented; truths have to be searched for, seen, realized and pointed out.

But in time, there occurs a need to change the language of utterance. There is a need to modernize understanding.

Three Human Brains
(Cranial, Heart and Gut)

In the field of science, an understanding is only valid till another understanding comes to replace it.

The new discoveries, with regard to the anatomy of the human body, lead to an important question. What exactly is the human brain?

We are conditioned to think that the brain means the mass of neural matter that we have inside our heads. The truth is something very different, something that was always known through the 'sense' in the heart and the 'feeling' in the gut.

Let's begin with the heart. It used to be viewed as a pump, a muscle called myocardium, located in the centre of the chest. But now it has been discovered that the heart has neural cells or 'brain' cells, and the neural control of the heart, or the 'heart-brain', has led to a new discipline in science, called *neurocardiology*, whose aim is to understand the intrinsic cardio-nervous system.

It seems that the 'heart-brain' makes functional decisions independent of the cranial brain (the traditional brain in the head), and radiates a powerful electromagnetic field that can be detected by scientific instruments from a distance of five feet or more.

The research is still at a nascent stage, its entire implications not yet fully comprehended.

But we also have a third brain, that hides in the gut, or the 'enteric nervous system', that has led to another new field in science called *neurogastroenterology*.

All the three 'brains' – cranial, heart and gut – are anatomically connected with the central nervous system.

Or in other words, the 'human brain' consists of three brains – the cranial-brain, the heart-brain and the gut-brain, which are all interconnected through the spinal column, nerves and the nerve centres.

So now, when the question arises as to what is the human brain, then one has to think, the entire human torso – from head to gut – (apart from the four limbs) is the 'human brain'.

Within this anatomy of the 'human brain' – the human torso – also lies the anatomy of the human self.

Anatomy of Self/The Skeletal Invisible

MIND (CRANIAL INTELLIGENCE)
(Seat of Individuality – I in terms of the Ego and Soul.
Propensities, interests, mental qualities,
dispositions, intellect, memory)

⇧   ⇧

HEART (HEART INTELLIGENCE)
(Seat of Soul Mind desiring *being-well*,
creates soul consciousness.
Soul-interest. To give, to realize/actualize.
Finer Feelings – love, compassion, altruistic affection,
impartiality, righteousness and wisdom)

⇧   ⇧

BASE (GUT INTELLIGENCE)
(Seat of Ego Mind, desiring *well-being*,
creates ego consciousness.
Self-interest. *To get, to acquire.*
Coarse emotions – hatred, anger, attachment,
pride, lust, greed, jealousy and ignorance)

⇧

NEUTRAL SELF ENERGY/CONSCIOUSNESS
(The fundamental instincts of self-preservation
and procreation)

(From the point of experience, the left side is the field of
the Ego Mind, and the right side the field of Soul Mind.)

The Root of Humanity – Conscience/The Moral Sense

The conflict between Ego Mind (*self-interest – ego-sense – well-being*) and Soul Mind (*soul-interest – soul-sense – being-well*) produces *conscience* or *moral sense* – the root of ethics, fairness and judgement.

The Evolving Self

The *four-fold action* – of feelings/emotions, of thoughts, of speech and of body – influences and evolves the *sphere* of self, moulds understanding and modifies response.

The self evolves through itself, with the experiences of its own self experiencing life.

The evolution of self is an innate design of life.

An individual self evolves by decreasing the influence of the ego-self (along with its negative emotions) and simultaneously increasing the influence of the soul-self (along with its positive emotions) within the inner sphere of the individual self.

The Meaning of Rationality

The meaning of rationality cannot be aligned only with self-interest.

Being is a blend of well-being and of being-well – of pleasure and of happiness, of prosperity and of wisdom.

The rational self has the broader intellect and understands the need for striking a balance.

The rational self understands that one has to put soul-sense before ego-sense, fairness before profit, being-well before well-being.

The rational self understands that one has to put soul-self before ego-self.

This is true humane rationality – the real, the essential and the vital.

Happiness (for the ego and for the soul)

Human beings have inherited the ego and the soul. They have separate minds of their own. And the meaning of happiness for the ego and the soul are also different.

For Ego

Happiness is doing what one likes.
Happiness is fun.
Happiness is companionship.
Happiness is pleasure.
Happiness is victory.
Happiness is power.
Happiness is status.
Happiness is wealth.
Happiness is well-being.

For Soul

Happiness is doing what one loves.
Happiness is love.
Happiness is soulful response.
Happiness is humane conduct.

Happiness is being ethical.
Happiness is doing good.
Happiness is spiritual realizations.
Happiness is freedom from ego.
Happiness is tranquility and wisdom.
Happiness is being-well.

In life, one needs a balance of the two. But the happiness of the soul carries more grains of joy and feels more real, long-lasting and better.

Conclusion

During a communal riot in western India, a man had cut open a foetus from the womb of a pregnant woman. Some weeks later, the man had realized what evil he had committed. Incited by anger and hatred, he had lost the ability to judge his own action, and had joined a violent mob to commit a trail of heinous crimes.

But the soul-sense that allows a person to judge oneself, the voice of conscience, had returned to him.

He repented, suffered and committed suicide.

The story is like a parable.

It is due to the force of the *Ego Mind* whose narrow perspective churns ego-consciousness (emotions of anger, hatred, pride, lust, greed, attachment, jealousy and ignorance), that all human crimes and acts of evil and corruption are committed.

But there is the opposing the force of the *Soul Mind* with a broader perspective that churns soul-consciousness (love, compassion, altruistic affection, righteousness, impartiality and wisdom).

This is the natural design of the human self. The motive of nature, of evolution, of the universe or of God (however one looks at it) is to lead a human being to the twin goals of well-being and of being-well.

What one does, how one acts, is the free will that is granted to us all.

A balance between well-being and being-well must be consciously sought during our brief lives on this blue planet circling a small star in the vast and enigmatic universe.

3

All great truths only occur in solitude, in the twilight moments of calm and of reflection.

It was late afternoon, the poet was at the window. He was thinking about the treatise that he had written, when a word appeared in his mind and caught his awareness. The word was 'revolution'.

He thought: what does the word imply? It takes a year for the Earth to complete a revolution around the sun, it takes millions of years for the sun to complete a revolution of the galaxy.

On the cosmic scale, revolution means completion of a journey.

On the human level, the meaning is not that different. Revolution means both death and rebirth, an end and a beginning.

It means the end of the old, the beginning of the new. It means change of authority.

The poet realized, an inner revolution had already occurred within him – the ruling authority had been dethroned and replaced.

Ego-mind no longer dominated his will.

But the battle had not ended. He had to remain forever vigilant, for the ego mind tries to fight back to take over its dominance over the human self.

He would have to play the twin role of a master and a subordinate – he would have to rule over his left stream, and allow himself to be governed by his right.

That's the right way of action – the soulful, the wise and the humane.

And now he clearly understood that the greatest enemy of the human self is a part of its own self.

His revolt would be to lessen the effect of the ego-mind in himself, and stand up against its various unjust manifestations which populate his immediate sphere and the world.

Then he thought that the lessening of the dominance of the selfish ego-mind with its narrow perspective along with the simultaneous rise of the humane soul-mind with its broader perspective is the only change that can help mankind solve problems, protect the planet and accelerate holistic development.

And the transition and the spiritual shift from the *human* to the *humane* had already begun, and would only gather more strength in time.

Those are the right words, the poet thought, *from the human to the humane*.

At that very moment, a gecko on the wall, hidden behind a picture frame, sounded an approval, not once but thrice.

It was a boyhood myth – whenever a gecko cries thrice after something has been said or thought, it is a confirmation of the thought as a truth.

The smile of the poet appeared on the reflection in the glass. His eyes strayed to the horizon. He could see the metropolis, the blue sky and the few easy clouds which journeyed with the wind.

At that moment, poetry occurred to him, and he went to his desk and wrote down the following words:

The Self

The two opposing streams,
within which the distinct 'I' –
an evolving symphony,
that is played by time,
and modified,
by one's own response –
of feelings, thoughts,
speech and actions.

And then,
beyond the duality of self,
the non-dual –
the void, the source.

This is the field of experience.
A self experiencing itself,
while experiencing its world.

4

The word *satya* is translated into English as truth. But satya isn't just truth, it is something deeper and broader than truth.

Satya is timeless truth, an aspect of hidden reality, a spiritual realization, a philosophical understanding.

One can weave many more meanings into it, but satya is closely related, in spirit, to the Vedic word *rta*, that means the way of things, the nature of design, an underlying order.

But the essence of satya doesn't lie in the understanding of a truth, but its *experience*.

It is not merely an intellectual understanding, but an *experiential realism*.

Whatever one may experience, it's only an experience of oneself. Different experiences can only trigger different aspects of our own self.

Satya, in its deepest sense, is the reality of the spiritual – a wondrous feel of oneself, an encounter, a realization and an effect.

5

The poet read through the words of his poem, The Self, a few times, smiled and felt a happiness in his heart. It had been a worthwhile effort to understand the self. He was more aware and clearer than ever before. His sense of being-well had been enhanced.

At that very moment, he felt the touch of truth. He felt a wave of ease and of lightness. A buoyancy spread within him. His awareness gained more clarity. He felt calm, tranquil and deep.

Poetry occurred to him once again.

> I am someone and no-one,
> one and the many,
> at different moments
> of time.

6

The poet was feeling the effect of his verse when the words 'no-one' got stuck with him.

Something happened to him. He walked to his desk, opened his notebook and stared for a while at the blank white page.

Then he could find the words to convey Satya, not merely an intellectual understanding, but the realized experience of truth.

> Beyond the self, lies my true face.
> Beyond the distinct 'I' of myself,
> Beyond the twin streams,
> Is the clear unblinking awareness,
> The unchanging self, the void within,
> An old feeling,
> No one and nothing,
> Forever nameless,
> A face without a face.

The Wheel

1

The distance between the poet's apartment and Café Good Hope was two cigarettes long. It took him half an hour to reach the café.

The poet stopped, almost habitually, in front of a newsstand. He asked for the magazine edited by his grandfather. He was told that the special issue would only be available after a week. He bought a weekly news magazine instead.

The café was, like always, almost empty. The poet occupied his usual table, facing the street, ordered his tea and lit a cigarette.

He had been patronizing the café for over three years. He liked the simple pleasures of the old-fashioned café, abandoned by the gods of change, a place where time itself had fallen in love with nostalgia and refused to contaminate the slow olden air with the manufactured smell of modernity. From the busy street, the plain building was an eyesore. It reflected the stubborn withdrawal of an old man who refused to mix and gel with the new trends of neon.

It was the very lack of gloss that had attracted the poet. He valued the atmosphere. It was the only public place that could offer him a sense of private space.

Most of the customers were regulars, who occupied their own tables, ordered what they always ordered and did what they always did – smoked, chatted, read newspapers and sat silently. The café never ran out of empty tables and stood for everything that was now passé. The two waiters had a bored look on their faces. Those who arrived to remain for long hours seldom left generous tips. Sometimes a family of fat cats brushed the legs of the customers who had grown accustomed to them. The waiters spoke to the cats in slang and often kicked them away. The menu had few temptations to offer. The poet liked a peculiar flavour of tea that was served with locally made biscuits, which made hard noises when one bit into them. The sound complemented the turning of the heavy iron fans which forever threatened to crash upon the dusty floor. Once an antique dealer had arrived with an offer to replace the fans with new ones. But he was shooed away by the café's owner, an old lady, who was proud of every bit that she had inherited from her forefathers. 'My nephew, I don't trust him. He says this building is falling apart, there is hardly any income from the café. But the land is prime, he says. The idiot wants me to sell everything. He says that the café is too dated for today's lifestyle. He doesn't care for sentiments. He says you cannot feed upon sentiments. But can you ever live without sentiments? What is there to live with anyway? I know that idiot will sell off everything and run away with some girl,' the old lady had told the poet. She suffered from poor health and often the nephew arrived to sit upon the cashier's chair and spent the whole afternoon making phone calls. The poet could see in his eyes that he was only waiting for his aunt to die, to be eventually carried to a height – the Tower of Silence and left for the vultures to devour her remains.

2

An hour passed easily in the café. When the poet finished reading the magazine, he realized it was raining outside. He ordered another cup of tea and watched the rain through the glass.

When the rain slowed to a drizzle, the poet walked back to his home. With the habitual movements of his reflex, he touched the elevator button, the dimpled key, the door latch and a few switches. Finally the cold water that he splashed on his face made him realize that he was back in his apartment.

After a while, the poet went to the window and slid open the glass. He saw the sky crack and light up. A moment later, a deep grumbling travelled across the horizon. Flashes continued to occur at regular intervals. Then a sharp lightning occurred and discharged a bolt of fear and awe. Taken aback, the poet closed the window and retreated.

Half an hour later, he washed himself, changed his clothes, made tea and ate a cheese sandwich. Then he changed the compact disc, programmed a selection, pressed repeat, took up his favourite position on the two-seater sofa and lit a cigarette. The Vajra chants set in symphonic music matured. The deep voice of the Tibetan master penetrated his skin. With every exhalation, he watched the strange forms created by strings of smoke. The ceiling fan was on – the air in the room was mildly agitated. The patterns of smoke swirled, danced and dissolved into a mild smell.

The cosy refuge forced the world to lose its grip on him. He felt peaceful and calm. Awareness of his own self – his unhurried breathing, the rise and fall of his chest and muscles of his lower abdomen – gained rhythm and prominence. The chants receded to the background of his awareness, the noisy plastic that had come off a cigarette pack, stilled itself in a corner. Even the monotone of the ceiling fan that persisted without music, became soundless.

3

At nine, the poet thought about dinner. There was cooked food in the fridge and a drawer full of home-delivery leaflets, but the poet wanted to go out for a drive. It was Saturday, most of the restaurants would be crowded, and the poet didn't like eating alone surrounded by the sociable murmur of a crowded space. He thought of packing some Indian-Chinese from a take away joint.

The car park was in the basement of the high-rise building. It was an area neglected by Maintenance – many of the fused bulbs hadn't been replaced, and the wet weather had aggravated the smell of oil and dampness. The poet negotiated the basement floor – crisscrossed with frothy trails of oil and water – and reached his car.

The drive took twenty minutes. The rain had stopped and the sky had cleared. It helped the poet. The parking was full. He had to park the car two blocks away and take a ten minute walk to reach the take away.

After he had collected his food, the poet realized that a spectacle was taking place in the sky.

It was a lunar eclipse. People in the street had stopped to marvel at the sight.

The shadow of the earth was gradually obscuring the moon. The moonlight that enshrouded the streets was slowly disappearing, corners were being overcome by darkness and the sky was changing its hue.

4

'I had thought time would heal but time only destroyed,' the poet's mother had told the adolescent poet during the time of the divorce.

The marriage between his parents had begun to disintegrate after the birth of the poet.

As far the poet could remember, the atmosphere at home was marked by tireless conflicts between his parents. Both of them were temperamental, easily excitable, vulnerable to melodrama and overreactions.

The conflicts usually began from trivial incidents and irritations (causes not explicit), grew into sarcastic comments and criticism, and finally exploded as a burst of stale acts whose only aspects were anger and frustration (in the form of abuses and loudness), blame and accusation (which were governed by the word 'you' and very rarely 'why'), threats (of divorce and suicide), actions and reactions (physical abuse, grabbing of kitchen knife, slamming of doors, etc.), intolerance (of each other), tolerance (because of the children's future), questions (very seldom to each other, never to themselves, and more often to God and to fate) and a final declaration that they were also humans and had their own individual rights!

Torn clothes, mild bruises, broken glass or crystal-ware and a doctor's prescription (aimed at normalizing blood pressure and to induce sleep, after animated complaints of 'severe' pain in the chest) were the sad and usual remains of yet another night of drama, which the audience – the poet, his brother, the maid and the cook – could never appreciate.

The after-effects of such conflicts lingered on for days, which were marked by refusal of food, sleeping in separate rooms, and giving each other menacing looks and spiteful glances. Even their silence had the feel of a sneer.

The good spells – social occasions, holidays, visits to the club, cinema and restaurants – were contaminated by underlying tensions. Anything could happen anytime, and anywhere.

Such conflicts always left the poet's brother in shock and distress. The assaulting screams spilled over him like a fierce net. While the battles raged, he trembled like a trapped animal waiting for something worse to happen.

But somehow the poet could cope with it in a way better than his elder brother. The sudden eruption of voices could only make his heart beat faster for a few moments, after which the fleeting anxiety didn't transform into anguish neither did it mark the arrival of gloom.

(The poet had developed a severity measure – whenever the noise would cross a certain permissible decibel level he would run into the room of battle, play his part like a boxing referee and attempt to separate his possessed parents from causing major damage to each other. And his parents also made it a point to start their bouts only when the young referee was around.)

Unlike his brother, the poet also never developed the habit of biting and chewing the ends of his pens and pencils.

Only once, in a surge of anger, when things had become unbearable, the poet had stormed out of the house to wander in faraway streets before returning late at night to find a police jeep parked outside their house. The policeman was a jovial man. The poet was considered too young to have run away with a girl. 'Did you fail in your exams?' he inquired without any unfeeling harshness. 'No, he is very good in studies,' the poet's mother answered for him. 'Then, why did you leave home without telling anyone?' asked the policeman. 'Just like that,' replied the poet without a grain of fright. He didn't suffer from any guilt for making his parents so anxious – he had paid back some of the anxiety that his parents gifted him in abundance.

'Don't do this again,' the policeman warned him mildly and proceeded to give a lecture that the poet heard without listening.

For a moment, the poet had thought of handing out to the policeman a complaint letter that he had written against his parents for making his life miserable. Probably the police could force them to improve their conduct, treat each other with tolerance and create a better atmosphere at home. The document was in a drawer of his desk. He had revised it four times. To empty his heartfelt grievances in a complaint letter and to actually hand it over to a policeman are two different things. The poet chose not to do the latter. He was prevented, not by fear, but by sympathy.

The poet's parents were relieved to see their son safely back home without having confronted any of the terrible adversities which they had imagined.

But in certain circumstances, the lurking dangers of the streets are less bothersome than the lurking dangers at home.

After everyone had left, the poet's parents started another violent altercation by blaming each other for the poet's act.

4.1

It is usually during boyhood that one begins to confront the problems of adulthood.

During the tiresome rifts, the poet had often overheard his father accusing his mother for 'lacking in depth and maturity' and his mother also retorting the same.

Both his parents began to accuse each other for not being the person they had fallen in love with.

It seemed strange to the poet that the people, who once married out of love, could discover so many factors of incompatibility.

Only later did he come to realize that people keep evolving throughout their lives – one who is loved at one point of time, may not be loved at another.

(People evolve through understanding – some evolve slowly and some evolve much faster. And those without conscience don't evolve, remain the same, and even decline.)

When the growth pattern between couples becomes uneven, many deep differences – of values, of habits, of purpose – tend to arise.

And when deep differences arise, how one reacts to them makes all the difference.

The poet's parents reacted to their differences with anger and ego. (When the ego-sense flares up, it tends to cause myopia, motivate rough actions, while anger shuts down sense, judgement and understanding.)

The poet's parents became a victim of their own selves. They allowed themselves to be governed by their ego-mind and coarse emotions. They behaved like sworn enemies, and tried to humiliate each other in front of their own children. They searched for faults in each other, rather than within themselves.

Such an attitude, governed by hate and vengeance, ruined every chance of a peaceful coexistence.

As time passed, the poet's parents continued to take turns to play the role of the oppressor and the oppressed. The shouting, the screaming and the madness became more intense. Every new event touched the lowest of lows. Instead of four glasses, five were broken, instead of three bruises, there were four.

The limitless decline could have only ended in murder or suicide. But fortunately for everyone, the marriage self-destructed; the poet's mother finally found evidence of his father's infidelity and couldn't bear the humiliation and the anguish.

4.2

The poet's father was involved in an affair with a female employee in his office. The woman was shrewd, manipulative and ugly. But often an unattractive woman is more adventurous in matters of sexuality than a beautiful woman. She tries to nullify her lack of good looks by being more open, raunchy and explicit.

The poet's father got hooked to the pleasurable diversion. His mistress was keen on kinds of sex that his wife viewed as dirty, vulgar and indecent.

The poet's mother had intuited something from the changes in her husband's behaviour. The choice of his clothes had changed, his mobile remained switched off at odd hours and his 'business' trips had increased.

Then some rumours started to circulate through his office employees. But the poet's father dismissed all rumours as absolute lies, lowly acts of his office employees, and accused his wife of being unfairly suspicious, and even crazy.

'Take your mother to the department of neuropsychiatry,' he often told the poet.

This had gone on for several months. The poet's mother became a detective and her bored friends gladly snooped around for her.

Then something happened, the poet didn't know how, but evidence was found. His father had been having an affair for over two years and had been lying blatantly to everyone.

A tedious melodrama erupted in the house. The poet's mother reacted with tears and anger. Friends and family got involved. Blames and counter-blames were exchanged.

During all this, the poet's father refused to show any sign of remorse or regret. On the contrary, he justified his behaviour by accusing his wife of being 'a good mother but the worst wife'.

The statement wasn't conclusive. But it was enough to suggest that he was unhappy with his conjugal life and was forced to stray elsewhere in search of pleasure.

In his mind he was no longer a husband – a father of two boys – who had cheated on his wife and had lied to his family, but a sad and poor victim of his wife's lack of libido, or sensuous finesse.

It never occurred to him whether he himself appealed to his wife. He only cared for himself and was prepared to say and do anything to serve and protect his interest.

He tried his best to change the picture and blamed his wife for all his actions.

'It is because of you all this has happened,' he shouted, 'not because of me.'

4.3

The selfish behaviour of his father looked absurd to the poet. He felt no shame or regret. The poet's mother could have given him another chance, if he had shown some remorse. But he continued to blame her, and that inflamed her further.

During all this, the poet's brother supported his father; he thought that his mother was overreacting and didn't consider his father's blemish as a major offence. His stand suggested, whatever his father did outside the house shouldn't have been too much of a concern. He said most of his friends reported that their mothers preferred the peaceful continuity of family life and turned a blind eye to the amorous diversions of their fathers.

'They have to do so because they cannot pursue independent lives. They are weak and dependent,' the poet had voiced, 'but mother has a strong character, a good job, and self-respect.'

But self-respect isn't necessarily a virtue to everyone. The poet's maternal grandmother, who always expected to be applauded for the sufferings she had borne as a wife, advised her daughter to let the season pass. 'Everything will be forgotten in time,' she had said.

But sadly, time doesn't have a weak memory. Deep grievances persist as shadows, old issues become the swords of new battles.

And the cycle of a good spell and then the unbearable, continues forever, without coming to an end.

4.4

The situation in the poet's home had hit a complete deadlock. The poet's father didn't move from his stand and the poet's mother didn't know what she really wanted. The only things that existed were suffering, anxieties and tensions.

But the poet knew something had to be done against suffering, some action had to be taken.

One grim evening, the poet suggested to his mother, 'Why can't we go away from here? You don't have to bear with everything. It's not right.'

The poet's words made his mother's decision to file for a divorce easier. To bear suffering and swallow humiliation were no longer the duties of a woman; inaction against unfairness was no longer a good character certificate. It was time to revolt, shun impotency and take action.

The poet's mother refused to be weakened by the few beautiful memories she had shared with her husband. She gathered her strength and filed for a divorce.

4.5

The decision spared the family from bitter conflicts and unbearable nights of melodrama.

It was a belief of the poet that it is better to separate than to suffer.

His parents' divorce didn't scar him, on the contrary, he felt the relief of finally waking up from a morbid nightmare. The only person whose absence mattered to him, his mother, was to remain with him. Others, to his own surprise, didn't matter much.

The poet's mother was the manager of a well-known travel company. She didn't want to remain in the same city where her ex-husband would live and move around.

The poet didn't revolt against the idea of living with his maternal grandparents like his mother did, but he understood when his mother explained why she wished to settle in a new city. 'I just want to get away from here. I am tired of the same old people, the same old things. I need some breathing space. We both can have a fresh start together in a new environment.'

She took a transfer to the head office in a different metropolis where she also had friends. One of them, whose husband owned a flourishing construction firm, helped her to purchase an upmarket apartment at an honest price.

Finance wasn't a hindrance; his mother's savings, the divorce settlement and his grandfather's generous contribution combined together to meet their needs.

The two and a half hour flight between the two cities covered the distance of years. It was still very difficult for the poet's mother to come to terms with the facts – over two decades of married life was over, the family divided, destinies separated.

And the nostalgia for a home that wasn't hers anymore drowned her in bitterness.

She tried to nibble from the tray that the flight attendant had brought, but she couldn't finish the food.

She fell back into reverie and sighed with heaviness, 'I really cannot understand how all this happened, and why.'

The poet got infected with sadness and gloom. But he refused to surrender, give in and relent. He revived himself by changing his thoughts. He got his mother to come out of herself and tell him about the flat she had bought, the metropolis that would now be their new home, the high-school where he intended to be admitted to and such other things. The flight, then onwards, became a journey.

4.6

After arrival it was a busy period. Many things had to be bought, many things had to be done, the poet had to carefully choose his subjects and begin his high-school life. He had passed his secondary school examination with a great percentage that had given his mother a new high that had lasted for four days.

After a month, when the rush of things had gotten over, a certain dullness surfaced in the evenings. There were less things to do than before, the evenings appeared motionless, long and oppressive.

But the mother and the son did their best to keep themselves in good cheer. 'Play some music,' the poet's mother would say, 'something nice and slow.' And before the poet rushed to get a CD, he would say, 'Make some snacks, something nice and quick.'

Between sudden laughs and spells of silence, the mother and son got along well. They argued over simple things – from the type of vegetables which were to be cooked more often to the kind of paintings which were to be hung on the living room walls. The hangover of the crisis bonded them closely. They discovered each other, not solely as mother and son, but also as individuals and friends.

The poet realized that the friendship of his mother wasn't her devious ploy to know about all his doings. She genuinely believed that her relationship with her son must be frank and candid. She had told the poet that when she was young her parents had never been her friends. She could never trust them and had suffered because of the distance. She wanted to become a friend whom her son could confide in.

'I am trying to rectify the mistake of my own parents,' she had told the poet, and the poet had thought, 'This is how progress is made. Every generation longs to rectify the mistakes of the previous.'

As days passed, the interaction between the poet and his mother became more frank. The poet got to know of deep truths which lay buried in his mother's heart. 'I married because of a mood swing. It was a sudden infatuation. I was young, naïve and sentimental. If your grandfather was a bit more understanding, I wouldn't have married your father without knowing him a bit better.'

'What had happened?' the poet asked.

The poet's mother took a deep breath, reflected for a while and then spoke freely, 'After college, as you know, I had got a scholarship. I went abroad. But I hated living in a cold unfamiliar country. I lost my interest to pursue the foreign language and become a teacher. I came back. That time I wasn't going through a good phase. I didn't know what to do, I was lonely, vulnerable and unhappy. Then I met your father. At that age you get taken in by appearances. He was religious. I don't know what that meant, but I was attracted to him. He gave me the patient hearing that I needed badly. He also got me interested in travel and tourism. But when your grandfather got to know of him, he made a big scene. Without his wish nothing happened in the house. I took your father to him. But then your grandfather told me not to see him anymore.'

'Why?'

'I don't know, he was never free with me. He expected that I should only obey him just like your grandmother. I hated the atmosphere at home, I just wanted to run away.'

'So a marriage to father was the best option,' the poet finished for his mother.

'See how well you understand. You have turned out much better than I ever hoped. A colleague at the office told me, children who witness bad marriages are wounded forever. But I told him, my youngest son has matured faster. Every passing day you strengthen my belief.'

4.7

Things went fine for several months. The poet's mother was generally happy at the travel house and the poet kept himself occupied with classes and extracurricular activities. Then another difficult phase began. The poet's mother was diagnosed with diabetes; she began to take medicines; kept forgetting where she placed her keys; seemed to be hurt very easily and began to accuse the poet of being too involved in his world to care for his mother.

Frequent squabbles, about when to come back home, late night parties, too many phone calls from girls, discovery of cigarettes in the poet's room, began to take place.

The poet's mother had always had hypertension – she had been doing far too many things throughout her married life which she didn't like or appreciate. Living with a husband who always incited anger and frustration was the chief culprit. The terrible discontent of having to act perennially against her own will had no way to surface except through high blood pressure. She had to be prescribed routine medicines, anti-anxiety tablets and sleeping pills.

The poet was habituated to his mother's emotional outbursts. He knew his mother would get angry, overreact and make a tiger out of a catfish. All he had to do was to prevent her from flaring up. So he guarded his own response, kept calm, avoided arguments and waited for the storm to subside.

But with the advent of diabetes, the poet's mother found it more difficult to cope with herself. She kept brooding over trivial things, sat on the side of the couch with a disgusted look, screamed unfairly at the maid and the driver, quarrelled with her friends over the phone and snored fiercely in her sleep.

All her daily dosage of medicines seemed to have suddenly failed. A sporadic tempest began to rage within her and forced her towards sudden outbursts of madness and anger.

She would make long-distance calls to the poet's father, from the landline, and ask him mockingly whether his lover yielded to his perversions, and then she would fire a volley of unspeakable abuses.

And whenever the poet would try to prevent her from making those calls she would viciously scream at him – a hideous part of herself would take the chance to erupt through her angry mind, burning with hatred.

The only 'mistake' of the poet was to say the truth – his father no longer mattered to them, so his mother shouldn't bother about him at all.

But the poet's mother somehow interpreted the poet's statement as something like 'let him enjoy his own life in peace'. Her reaction seemed to suggest that the poet was guilty of wanting to protect the peace of his father rather than do something drastic to avenge his mother's humiliation. As if it was his 'duty' to prevent his father from sleeping with his lover, take a flight back to bang his father's head, cage him with a chastity belt and bring back the keys to his mother.

As if such extreme behaviour was naturally expected of a true son, and by forgetting his father, the poet was proving that he had no reason to punish his father and had pardoned the man.

This perceived inaction was unacceptable to the poet's mother. She would flare up fiercely spitting venom at the poet with thoughtless statements – 'You have his bad blood running in your veins, how will you understand my pain?' – that deeply hurt and angered the poet. He disliked his father and wanted to be his opposite. The matter of 'bad blood' – his compulsion, not his choice was an unfair thing to say.

Once the poet had furiously protested – he had to – and his mother had broken down in tears. And after a while, she found the poet in his room and apologized for her behaviour. 'When I was young I was not like this. I never had a bad temper. My life has only brought out the worst in me,' she had said in a tone that had touched the poet's heart. He could never remain angry with her for more than a few hours at the most. 'Why do you behave like this? I really cannot understand,' the poet said. 'You do so well in your work, everything is going so smoothly. You don't have to be so angry and so negative, you only poison yourself.'

But again at the dinner table, while nibbling at her plate, his mother would sigh with vacant eyes, 'Everything is finished, everything.'

4.8

In the behaviour of his mother, the poet closely witnessed both – the mayhem of a mind ridden with anger and the sad perils of attachment. Both conditions made him cringe in fear – he had to avoid them in his life at all costs. Attachment to someone is natural, but if that leads to an obsession, then it becomes enslavement.

His mother's revolt against her husband by slapping him with a divorce had failed to free her, she was still enslaved to the man. If only she could succeed to root him out of her life, and learn to view his memory with scorn and indifference, only then would her revolt succeed, only then would she emerge triumphant. The poet had understood this, but his mother hadn't. She continued to do what was the easiest thing to do – to lose self-control and harm herself.

Trying to manage and discipline his mother had become a full-time pursuit for the poet. He inspired himself with his own youthful slogans (ideas have to be transformed into endeavours; anguish will have to be acted upon) and realized he had to take an initiative, do something.

He thought of ways to calm his mother down. Then he realized that guidance should come from within. So he bought self-help books and spiritual scriptures and gave them to her.

But after a few days, she became more melancholic. 'I cannot read them any longer,' she had told the poet, 'they remind me too much of my own mistakes.'

But the poet refused to give up. He employed other methods – he placed a laughing Buddha in his mother's bedroom, lit a green-coloured bulb to soothe her nerves, played meditative music and spiritual chants to create a peaceful atmosphere and slipped books of wisdom under her pillow so that his sleeping mother could somehow be influenced by them.

The poet's tireless efforts did manage to cheer her up. But whenever she spoke to the poet in one of her calm moods, a theme of finality began to lurk beneath her thoughts. 'When you were a child I used point to the stars and tell you, after people die they become stars in the sky and watch over their children forever.'

'I remember,' the poet said.

'After I die I want to be born as a star so that you can look at me always.'

The seventeen-year-old poet didn't like the idea – he couldn't imagine the absence of his mother. He also didn't like drowning himself in gloominess. He turned the conversation in another direction and implored his mother to make efforts to remain calm. 'Don't watch too much television and avoid making calls,' the poet told her. 'Sit in your room silently for a while.' But silence had a terrible effect upon her. Whenever she remained alone, disturbing feelings and memories overcame her, her blood pressure increased, she negated the instructions of the doctor and helped herself to her favourite chocolates.

A new sub-season began, the poet's mother began to suffer from sporadic spells of doldrums.

It wasn't as severe as clinical depression, it was a state of mind afflicted with numbness, vacuity and unhappiness.

Whenever his mother would tend to fall into her slow moods, the poet would try to make her remember the few happy memories and manage to make her laugh. (Memories of heart-warming incidents of one's children, especially when they were infants, are the best medicine for unhappy mothers.)

The poet had realized, depression is infectious; numbness spreads from one person to another. So he made his mother talk about his own antics as a child – how the baby poet had peddled away all the blankets and the sheets, how he had woken up at midnight and had wanted to play, and other such things.

The strategy of evoking happy remembrances was effective. Once she was in a lighter mood, she always loved two things – to cook great food, and to receive praises.

The poet soon realized what his mother loved most – she loved to surround herself with people. All women love a sense of community, she was no different.

The poet's mother tried to live sociably and that worked well for her. The weekends began to get filled with exotic recipes and a gathering of friends. Sometimes the poet's grandparents arrived to spend a few weeks with them and sometimes the poet's brother would make a phone call. The poet's brother was headed overseas for his education that was to be funded by his father. He was going to attend a second-grade American university to study business management. He didn't speak of any intent of ever coming back, even for a brief holiday.

During the time when the poet was about to take his final high school examination, the poet's mother had almost recovered. She was more calm than ever before and her emotional hangover had lessened. But petty differences with her son continued to surface.

(The poet pursued several extracurricular activities which made him appear as one who neglects academics. He played cricket for his school and the under-nineteen state team, and zealously participated in all school fests as an active member of various societies. A new shelf had to be put up in his room to accommodate all his prizes.)

The poet's mother wanted the poet to get good degrees which would have a high market value, a high-paying job, a professional wife and a couple of children. Quicker they arrived, the better. The poet's mother had already thought about his future and sealed his destiny and didn't want the poet to shatter her dreams.

The poet knew, when the time came, he would only do what he wanted; any amount of pressure or emotional blackmailing wouldn't manage to twist his will.

But the task to make his mother happy still burdened him. He absorbed himself in his own world of adolescent adventures; his mother filled the emptiness of her spare time with her friends, who arrived to spend long hours every weekend.

All the friends of the poet's mother were going through a lonely phase in their lives. The children were growing up fast, relatives were getting devious and troublesome, and their husbands were becoming indifferent and careless.

So they supported each other by forming, as they called it, 'a family of friends'.

Apart from their own lives, they spoke of other things such as clothes, exhibitions, food and gossip. They also laughed a lot. Their jokes always meant making fun of others.

The quartet of women got along fine. They didn't try to dominate each other, didn't try to show off, and maintained peace, goodwill and harmony.

4.9

Peer pressure works more in adults than in teenagers. The poet knew that his mother's friends also ingrained many fears into her. The most significant one was the fear of loneliness.

The poet understood that his mother was toying with the thought of a second marriage when he saw her surveying the Sunday matrimonial supplement.

'Did you find anyone interesting?' he had asked playfully. His mother had smiled and asked, 'Can you tell a person from five lines of print?'

The poet then jumped beside her and read out the profiles of the aspirants. 'Here's a suitable one,' the poet had read from the column of second marriage. 'Widower, 58 years, 5 feet 8 inches, children well-settled, businessman, several properties, seeks alliance with attractive liberal-minded lady aged about 45, apply with full-length photo.'

'Fifty-eight-year-old pervert!' the poet's mother had exclaimed. 'Apply with full-length photo. Rascal.'

The poet and mother had laughed and joked for a long while. She found all the profiles distasteful and repulsive.

But it became evident to the poet that his mother felt the need for another husband, or at least, a companion.

4.10

The poet's mother always wished that the poet should be on his own during his college years – it would teach him self-reliance, free him from the domestic atmosphere and allow him to devote himself completely to life in the university.

After the higher secondary results were declared, the poet found a place in various colleges, but chose a college that was in another city.

As the day for the poet's departure arrived, the poet's mother was engulfed in melancholy. She also told him that she was lending serious thoughts to a second marriage.

The idea of his mother's marriage didn't appeal to the poet.

It was also not clear to him, and perhaps not even to his mother, whether she really felt the need of a companion because of loneliness, or simply wanted to get another man to make her ex-husband jealous.

But whatever the real reason might have been, the truth was that she felt the need for another man.

Understanding the truth, the poet had told his mother that he had no problem if she decided upon another marriage.

'We should keep each other happy for our own sake,' the poet had said.

His mother had displayed deep relief; she had hugged and kissed him with all her deepest love.

After three months, the poet received an email from his mother that he read on his laptop in his room at the college hostel.

My dearest son,

I am very happy, extremely happy. I don't know how to tell you about it but I have to. Before I had known your father I had a close friend in my younger years. To speak frankly, I was quite fond of him. He was also very fond of me. We were very good friends. He wasn't like the usual young men we used to know during those days. He was very different. He used to read a lot, write poetry and told me many things which left a deep imprint. But circumstances made us drift away from each other. After that, I met your father and loved no other man since then.

But life can be so amazing. After twenty-six years, I met him a few days ago. I recognized him in a single moment. You can well imagine how I felt. To meet someone after so long, and so unexpectedly, I really couldn't believe it myself. We went out for dinner and talked a lot. He is still single. All his youthful interests still remain with him. He was delighted to know that you also read a lot and write poetry. He is a very nice man, life has only made him better. You remember the talk we had before you left? When we discussed how difficult it is for a woman to live a life without support? Of course, you are my life's support but you may have to go abroad, study and pursue your career. I will not have anyone to live with. God understood my distress. When I needed someone whom I can trust, he suddenly sent the most trusted one back into my life. We spoke of marriage, and agreed. I know your midterms are coming. I think it would be better if you meet him first. I know you will like him a lot. If you cannot come home for a few days, then we will wait for your summer vacations. I am feeling so happy that I cannot explain. Please write to me fast about your thoughts. And don't forget to eat breakfast. I read in the papers that you can skip lunch but cannot afford to skip breakfast. Breakfast is the most essential for keeping good health. Your exams are coming, you should not neglect your health. Do write fast and call me up when you have read the mail. Take care and be a good son, like always.

With lots and lots of love,
Your loving mother.

The poet's initial reaction was neither happy nor sad. Then a feeling resembling a removed sense of loneliness seeped within him. His mother suddenly appeared to have retreated to a distance.

After a while, the poet shrugged off the uneasiness and got his mind to reflect. Then he realized, his mother's impulsive decision to get married to his father was influenced by another reason: she had looked upon her marriage as a way to recover from a failed love affair.

Perhaps that's why she became more impatient and frustrated when her 'businessman husband' didn't resemble the traits of her 'poet lover'. Perhaps the sharp contrast between her husband and her lover had continued to haunt her.

Or perhaps, nothing of that sort took place, she had forgotten her lover, got on with her life and everything else had just happened.

The poet realized he would never know the truth for sure. And it was better and wiser not to speculate, or allow his imagination to run amok.

The poet's mother had already prepared him for the possibility that he would have a stepfather. Even though he wasn't comfortable with the idea, he had given his word of consent. And it made no difference to him whether the man was known or unknown to his mother.

But the former, a known person, was better for his mother, than the latter.

After rereading the email several times and smoking a couple of cigarettes, the poet felt he had to confront the new reality and speak to his mother.

She sounded so happy and so spirited that the poet had also cheered up. He told his mother not to wait for him and good-naturedly joked, 'He had drifted off once, don't let him drift away this time.'

His mother laughed with the blooming enthusiasm of a teenage girl. The poet understood the resurrected laughter of first love, ever fresh and ever promising. His mother was alive again. The catalyst – the comet of old love – had returned from oblivion, as a shining streak.

4.11

A month later, another email arrived from his mother. His mother informed him that the marriage registration was over. The poet had a new mentor. That's the word she used, 'mentor'. The poet suspected that the word had been suggested by his stepfather. He smiled to himself thinking that his own father had never been a mentor. He never taught him anything that was good and worthwhile.

The letter had also explained that the poet's mother wasn't changing her residence. She would continue to live in her own apartment and the poet's 'mentor', who had his own apartment, would keep shifting between the two. And this arrangement would suit everyone, including the poet.

After four more months, two months before the first year final examinations, the poet received a parcel and a note:

Your mother advised me not to send you new books. But at your age, I was always interested in reading everything except textbooks before any exam. Thus, I have taken the liberty to gift you these which will serve you their worth.

Your mother told me that you are taking a long holiday down south with your friends after your exam. I am looking forward to meeting you when the effect of the forests, the mist and the blue mountains will still be strong within you.

Till then, adieu.

5

The front page of the Sunday morning newspaper flashed a picture of the lunar eclipse.

The poet had woken up late. When he turned towards the clock, he realized he could sleep for another hour. But he dismissed his lazy impulse, got up from bed, found his slippers and put on some music. He drank his customary glass of water (kept overnight in a copper jug), brushed his teeth, ate a handful of walnuts, almonds and raisins, did his stretching exercises, relaxed his rhythm of breathing, prepared his cup of green tea and smoked his morning cigarette.

After a while, he opened the entrance door, picked up the heavy bundle of newspapers and kept them on the dining table.

Being a Sunday, the morning cook and the cleaning lady wouldn't be coming. So he prepared his breakfast – grilled egg sandwiches with cheese, apple and banana, and carried the plate to the table. (He liked to eat his breakfast while reading the morning newspapers.)

The poet subscribed to three English dailies. One of them carried the picture of the celestial event on its front page.

The sight of the eclipsed moon wasn't dramatic like a solar spectacle, it was more sublime and ominous – light didn't lessen but the darkness deepened; birds didn't head back to their nests, but became more silent in their sleep.

In one of the photographs, the shadow had progressed to cover a third of the moon, the rest of it was hazed by an eerie tinge of reddish brown. It was ghostly and supernatural. Unlike the solar phenomenon, the lunar eclipse didn't appeal to the feeling of magic, but stirred a sensation of unease and horror.

The poet decided to preserve the picture for posterity. He cut it out and thought of keeping it within the pages of a book. He decided against it and thought of preserving it inside his boxes of memorabilia.

The two cardboard boxes were kept in a chest of drawers. He retrieved the one that contained all the personal letters and cards he had ever received in his life. While keeping the picture, he came across the note that his stepfather had written to him seven years ago. He read the note twice and felt something like regret. The poet was supposed to meet the 'mentor' after his holiday. The well-preserved colonial bungalow of his college acquaintance, where they had stayed during the holiday, was on the slope of a blue mountain. The distinct murmur of an invisible stream reached the house from below. Amidst the orchestra of insects and guitars, the nights were cool and wild. The joker had got drunk and puked on the ornate furnishing of a periodic daybed. He felt so guilty that he got drunk the following night and puked on the antique carpet. The following morning he took a vow. He would never succumb to the black magic of rum. The vow had lasted eight hours.

4.12

The poet had known from the voice of his mother that something was deeply wrong. But his mother hadn't told him anything over the phone. After spending two weeks in the mountains the poet returned home. His mother looked like stone. She had shrunk and her complexion had darkened. It was obvious that she hadn't touched food for several days. The poet's arrival brought forth her tears. She told the poet that his stepfather was dead. Two months ago, his stepfather had left without letting his mother know about his whereabouts. For several weeks before his disappearance, he seemed to have been caught in a turmoil. He couldn't hide the disturbance from his wife, who remembered his traits which had persisted in him since his youth. Knowing him, she didn't press on with her questions, and left him alone. Then one evening when the poet's mother returned home, she noticed the absence of his clothes and travel bag. He hadn't left a note. His phone was switched off. But a day later, he wrote an email where he indicated that he was fine, asked her not to worry and told her to wait for further news. A month later, a friend of his stepfather arrived to let her know of his death. The poet's stepfather had discovered he had terminal cancer and couldn't bear to die in front of the poet's mother. He had gone to his friend who lived in a hill station, and within a fortnight he was dead. His final rites were performed by his friend. The ashes along with the mud vessel were immersed in the Ganges. After a couple of months, the heart of the poet's mother failed. She died peacefully in her sleep. She was a year younger than the poet's stepfather; he was forty-eight when he died.

6

After a week, it was Sunday again. Since the moment the poet had woken up, he had been thinking of his dream. He was with a woman, a stranger, with whom he travelled in a metro and visited an art gallery where he met his mother who was with another man, whom the poet now thought of as his stepfather. All of them entered a hall that was a tall library containing gigantic books which they could only gape at. Then, a clamour of temple bells sounded from somewhere, a man from the poet's office came up to them to sell digital watches, a few old lions silently walked past.

The poet thought of the dream all morning and dismissed it around noon. But he knew that this was a sign – his creative faculties worked more effortlessly when he dreamt absurd dreams. The poet realized it was a good time to write poetry. He lit a cigarette, went to his study and turned on the computer. He checked the feature article that he had written for the newspaper. He amplified and omitted. He took a printout; he knew he would be tempted to make more changes after a while. He decided to reread the article at night. He was about to turn off his computer when he remembered that he had to write to the joker who had mailed to invite him to the mountains and stay in one of the two recently renovated hotels; the joker had written that he had designed everything himself and the rooms now were cosier and more world class. The poet wrote back that he would make a trip very soon. Then he turned off the computer, ate his lunch, went to his bedroom and watched television. After completing three cycles of surfing, he watched the lifestyle of African elephants, highlights of Latin American soccer, discussions of British journalists, catwalks of French models and parts of an Indian movie.

After an hour, he forcibly switched off the television; he wanted to do something constructive. He wandered around the apartment with a glass of sea buckthorn juice.

The Sunday afternoon lay in front of him like a monsoon seashore, lazy and unpopulated. He decided to fill it with music and some reading.

The poet adjusted the volume and kept the music of *Sigur Ros* at the background of his thoughts. Then he realized he didn't have a book to read. He approached the bookshelves and surveyed the titles.

All the books fell under three categories – the books he hadn't read properly, the books he would never read again and the books he reread.

The poet decided to concentrate on the last category. He found a collection of essays on love. He realized that the book had arrived in the parcel that he had received from his stepfather seven years ago.

The poet had read twice the two novels and the collection of shorter works but he had read the essays on love only once. It was time to read the essay again. He leafed through the contents. There was an essay by Schopenhauer.

The poet went to the pages. The philosopher had quoted a few lines of a novelist whom the poet hadn't heard of – Mateo Alemán. But the quote from *Guzmán de Alfarache* got him to reflect. It said: 'For one to love it is not necessary that much time should pass, that he should set about reflecting and make a choice; but only at that first and only glance a certain correspondence and consonance should be encountered on both sides ... and to which a special influence of the stars generally impels.'

Few other essays in the book earned the poet's interest. But he didn't want to remain buried on the couch. The only other place he could think of going to was Café Good Hope.

It was the only public place in the metropolis, overrun by shopping malls and multiplexes, where he could peacefully read the book, drink tea and smoke.

7

The poet liked driving – it ignited his awareness. Whenever the poet felt lazy and dull, he went for a drive, it made him alert, and helped him to regain sharpness and focus.

The poet walked to his desk and picked up the car key that was always kept in a small yellow limestone bowl. But a thought made him pause. He put the key back in the bowl. Parking was a major problem near the café, so he decided to walk.

The walk took five minutes more than usual – the poet had stopped to buy a copy of the monthly astrological magazine edited by his grandfather.

Sitting at the deserted café, it took him less than fifteen minutes to read the few articles of his interest. He felt disappointed; he didn't learn anything new. He discarded the magazine, gazed through the glass and observed the passing of time.

The people on the pavement crisscrossed his vision. It was four-thirty on a Sunday afternoon. The traffic along the road was sparse. A hopeful boy waited for the lights to turn red and pleaded with the people in the cars to buy incense sticks. The boy never bothered those who walked on the pavement. Allured by the temptation of feeling the momentary whiff of the coolness of an air conditioner, he rushed to the luxurious cars and pressed himself against the cold glass, and hoped that the glass would slide down due to the benevolence of the motorists.

The poet returned to the book he had carried with him. He reread the essay by Schopenhauer. This time around, like always, he discovered fresh insights which modified his previous understanding of the work.

All great passions arise, as a rule, at the first glance, wrote the philosopher. Passions may be spontaneous, the poet thought, but love takes time to deepen with trust and respect.

After a while, he focused on the other essays. His concentration collapsed when he realized he had finished his pack of cigarettes. (The poet's greatest nightmare was to discover that he had run out of cigarettes in the middle of the night.) He gulped whatever remained in the tea cup, kept the book on the table, gave a signal to a waiter that he was going out to buy cigarettes.

When he came back, he saw a woman – clad in cargo pants, white top and a large leather bag on her left shoulders – standing by his table and flipping through his book.

The poet walked to the table. When the woman turned towards him, the poet realized, she was about his age. Her long straight hair was parted from the middle. Her attractive angular face – sharp and dusky with large eyes, had a look of soulful simplicity.

The poet somehow could *feel* her presence, and that arrested the focus of his attention.

The arrival of the poet conveyed the message to her that the owner of the book had returned to the table. In a tone of courteous apology, she said, 'I was just looking at your book. I hope you didn't mind.'

The poet said he didn't and offered her a seat. 'You can take a longer look at it if you are not in a hurry.' She smiled beautifully and drew out a chair. 'I just walked in here and was about to leave when I saw the book on the table,' she explained in a voice that suited her face. The poet became aware of her perfume. He also noticed the mythical motif of the Sun, hand-painted with bright fabric colours on her white top. She wore junk jewellery – earnings and bangles – which made tiny metallic sounds as she manoeuvred her handbag on the table.

The conversation flowed easily; this was the first meeting of these two strangers but it inspired an unusual confidence between them. As if they were two backpackers who had met at a café in a faraway land and were sharing their deepest thoughts and stories of life in their very first meeting, without any inhibition.

Both the poet and the young woman spoke of the book, and then about each other. She introduced herself as an artist, a painter.

'I started out on my own about two years back. I was in an art school before that.'

Then the conversation drifted to art, famous artists and the artistic trends of the time.

'Looking at you,' the poet remarked, 'it doesn't seem you would ever colour a painting to suit the living room walls of your collector.'

The poet's words surprised the artist. 'In fact, I strongly feel about this issue of turning art into decorative items and investment options. How did you know that?'

The poet smiled. But there was a brief silence, and the poet realized he had to say something more. 'You have a sense of idealism about you. And I also read people quickly,' he said.

The poet had sparked the artist's curiosity. She asked a flurry of questions and the poet answered, with brevity, precision and truthfulness.

If a man who has aroused the curiosity of a woman fails to maintain it, then the woman more often than not loses her interest in him. Women usually want to form a perception about a man very quickly, but some women do feel attracted to men whom they cannot easily understand. It is always favourable for a man to be in no hurry to let a woman understand him. Such a stratagem works as a form of subjective flirtation. Unlike sexual flirtation, whose motive is to arouse bodily desire by approaching each other with words and glances, subjective flirtation wishes to appeal to the other person's intelligence. It is sensuous not sexual, something that heightens interest, and gives rise to a pleasure derived from the responses of each other. But its primary motive is to arouse an interest to know the other as an individual. One seeks to get into the mind of the other and create a sort of cerebral intimacy. This flirting of minds outwits the flirting of senses. Not only does it possess more charm, promise and adventure, but it can also lead to the engaging beginnings of friendship, affection and even love.

The poet couldn't take his eyes off her. There was something about the artist. Her presence made him feel very calm and natural.

'This is a great book you have,' she said. 'All the major philosophies of love in a single collection.'

The poet could feel she was waiting for a response; he didn't want to disappoint her. 'Since I already know you will care to return the book, you may borrow it.'

'How can you be so sure?'

'I don't know, but that's the only way I can make sure that I have your phone number.'

The poet was relieved that she didn't blush, but she smiled beautifully.

'You want tea?' the poet asked.

'Yeah. I don't mind.'

Tea arrived in a few minutes. The conversation revolved around the book. The young woman said, 'I feel philosophers may want to understand the workings of love, but deep inside each one of us there is a longing that refuses to rationalize love. We are just happy to experience love, suffer from it, doubt it, disbelieve it, but somehow always hope for it.'

'But there is a sublime distinction between love and a lover,' the poet said. 'When someone says, I love you, it really means, you trigger the love I already hold in myself. And when someone falls out of love, it means, the lover no longer holds the power to effect the love that is ever present within the self.'

The artist thought for a moment. 'You are right,' she said. 'We love someone because the person allows us to have a deeper and a better experience of our own self.'

'And also,' the poet almost finished for her, 'the truth of our own inner self determines the depth of love we can possibly experience.'

'That's a nice thought,' the artist remarked with a look of reflection, that made her face look more beautiful. 'It also means that the depth of self-realization is the key to feeling, experiencing and understanding love.'

'Not only love,' the poet added, 'everything. Life itself.'

8

Six years ago, after the mourning period of the poet's mother ended, one late afternoon, a man in his late forties paid a visit to the poet. He said at the door that he was a friend of the poet's deceased stepfather and wished to meet the poet.

On the sofa in the living room, the visitor sat down with a mechanical poise, it appeared as if he was carrying a weight in his soul, a discomfort and a cloud of despair. The poet sat opposite him, on a chair, and waited for the visitor to speak. For the first couple of minutes there was silence; the lost gaze of the visitor was fixed on the carpet. Then he spoke very slowly, more to himself, than to the poet.

'I came here once before, to break the news of your stepfather's death to your mother. After that I used to call her up, now and then. Three weeks back I had called, your grandfather told me that she is no more. I was shattered. I didn't have any strength left in me. But I felt I had to meet you once,' the visitor said. Then he paused for a few moments; the whirring of the ceiling fan became distinct. The visitor continued to reflect silently.

The poet broke the uneasy silence magnified by the sound of the rotating blades of metal.

'Would you like to drink something?' he asked.

'Water would be fine.'

The poet got him a glass of cold water. The man drank it in a gulp, letting drops trickle down his chin. The water appeared to revive him. He exhaled deeply and looked at the poet.

'You are a brave young man,' the visitor said in a more firm voice, 'I can see composure in your face. Your stepfather had told me about you. He had said you also write poetry. He regretted that he couldn't meet you. He died very quickly, without suffering.'

Then the man suddenly got up and walked towards the poet who also had to get up. He held the poet's shoulders with his outstretched arms. 'Keep your strength. Your stepfather would have been proud of you. He was a spiritual person. He used to say, anguish is a friend who brings you closer to yourself.'

Before the man left, he had given the poet his card and assured the poet of any help he might ever need. The poet was tired of courtesy visits. His frame of mind was full of disquiet; he only wished to be left alone. Beneath his calm exterior, the aftershocks of his trauma, anger towards his father for having the audacity to ask his mistress to call him up and say 'now I am like your mother', and the strength that resisted both trauma and anger, were still engaged in a tiresome battle.

The poet misplaced the card of the visitor. His mind judged the visit as insignificant, and erased it from his memory.

8.1

The visitor – the friend of the poet's stepfather – stepped out into the street and looked up at the sky. A pensive evening was about to descend. A weak breeze was blowing against his face. He searched a part of the evening sky; he couldn't spot any circling bird. He remembered his friend – the poet's stepfather, and an old phrase that had stuck with him since his youth. He finally knew, Time had flown away.

When someone we love dies, with him or her, some fragment of our own identity also perishes. A secret face of ourselves might have only surfaced with someone, for whom, we were someone else.

With the disappearance of the loved one, a part of our own self also sinks and forever disappears.

At that moment, the man realized that an identity of him had perished. He remembered his impulse for laughter whenever he was with his friend. 'You are a clown,' his friend used to tell him. He smiled at a memory. Something swelled within him. But he also realized, he was no longer the joker.

The man hailed a cab. The taxi joined the traffic to take him to his hotel. Images and words kept occurring within his mind.

He remembered his friend whom he had met in their college hostel. His friend had a sweetheart. He had thought that they would get married. But they had separated. The sweetheart had gone abroad to study; she came back soon. But by then his friend had acquired a taste for solitude. The sweetheart married a man and moved away to another city. His friend had chosen to live alone, work as a journalist and pursue poetry. 'I turn the lights and fans on for company,' he used to joke. He was like that – a lover of distance. But then he again fell in love, with an artist, she was a spark of life. It lasted for seven years and ended in a tragedy. Fate had conspired to keep him solitary. He withdrew into himself, became a workaholic. Over the years, he published books of essays and poetry, travelled widely, became the arts and culture editor of a leading newspaper. And then, after many years, he met his sweetheart again. She was now divorced, a mother of two boys, and the youngest one wrote poetry. That had made his friend happy. 'The search will always continue,' his friend had told him with a spark in his voice. Then it happened like a fairy tale, after twenty-six years, the old college lovers got married. But time had run out. He came to him to die. He saw his friend dying in front of him; he couldn't do anything. In less than two months, the sweetheart also died. Their story of love finally ended.

The rotating wheel had completed a cycle. Now, today, he had come to meet the youngest son of the sweetheart, another poet.

8.2

On a radiant morning amidst snowy mountains, the poet's stepfather arrived at the house of his friend. He had an uneasy smile on his face. He looked tired and dishevelled. But the uncombed hair and the darkish stubble somehow magnified the shine in his eyes.

Later that night, he told his friend he had come to die in peace and leave instructions.

He struggled to breathe and spoke very slowly. He said that the lung cancer, instead of growing inwards, which would have led to lung abscess and the symptom of bloody coughs, had grown outwards. This allowed the cancer to infect the respiratory system in complete silence. He had no hope. The doctor said the case was peculiar – the condition was deteriorating very rapidly. The doctor wanted to experiment with the disease. He made arrangements to delay his death in some hospital. But he had told the doctor he didn't want an unnatural death.

Then he had explained he didn't wish to die in front of his sweetheart who was now his wife. She would have made him feel guilty for the fact that he was dying. She would have made him suffer from a pain that didn't belong to him. He only wished to die without drama, without disturbance, in silence and in peace.

He had already donated his books to a library, given his clothes to a charity, sold off his flat and deposited the money into the account of his wife. After his cremation, he instructed, his ashes must be scattered in the Ganges.

His condition worsened with every passing hour. But every evening he went for a tedious walk and watched the play of lights which crowned the twilight mountains.

And in the nights, he gaped at the blue-black cloak, studded with sparkling diamonds.

But on the eighth day, he refused to be tempted by nature. 'No more attachments,' he thought.

He preferred to remain alone in the room, stare out of the window and reflect.

Now, the world had simplified itself. There was nothing to oppose and to tackle. Every moment had become an occasion for aesthetics – churning of the soul consciousness.

8.3

Sometimes, one has to recover a bit from an illness, in order to die from it.

On the morning of the day he died, he looked less ill. But a few hours later he began to feel something. He realized, the wave had reached its shore; it was about to break, and become one with the ocean once again.

He lay on the bed and closed his eyes. He held his awareness just below his forehead. Soon his mind deepened, he was drenched with deep currents – lightness and peace held him in their womb.

Poetry hadn't deserted him. Play of words still offered him those moments which he also sensed in the beginnings of dawn, in the lips of a beloved, in music and in farewells.

His last verse was granted to him.

> At the end of all things,
> all is not sadness,
> a new journey will only begin,
> to know what lies beyond,
> dreamless sleep or another awakening!
> And the truth of the other shore,
> will finally, reveal its secret.

Then he recalled the amazing narratives of other worlds which tradition promised. Now he would know for sure, he thought, truths and falsities of afterlife or its absence – the dark river full of bones, astral doors through the heart of the sun, heaven and hell, the wheel of life, Elysian fields. A childlike wonder overcame him. Then a strange contraction grasped his body. He winced a bit, began to struggle. Links started to snap. His body started to feel heavy, his mouth started to feel dry. He began to feel cold, his breathing became laboured. And he felt as if he was being dropped into an expanse of space.

Images from his elapsed life pranced across his mind – his boyhood sparrows, the faces of two women, the sight of snow mountains, his mother, a path in a forest coloured with mist, the waters of a dark river...

9

In Café Good Hope, the young woman – the artist – was explaining something to the poet.

'If you hadn't carried this book with you then we perhaps wouldn't have met. When I walked in here, there was no one around. I was about to leave when I saw the book and the magazine on the table. Astrology doesn't interest me but the book caught my interest. *The Philosophy and Metaphysics of Love* is a curious title.'

The poet picked up the book. A sense of wonder overcame him. 'The book was sent to me nearly seven years ago. Somehow only today, I decided to read it, once again.'

'So you see, nothing happens by chance. All those chance happenings had a purpose,' she smiled. Her eyes sparkled with intelligence. 'Probably we were destined to meet here today. Perhaps, chance is a guise of fate.'

Then she laughed. The poet felt his heart stir. He knew he already liked the stranger. He spoke with a happiness that made the face before his eyes even happier. A face without a trace of make-up. A natural face of someone who also felt a similar connection.

'Perhaps it's true. Chance is not really chance, coincidence, not coincidence.'

The poet felt like smoking. He lighted a cigarette and exhaled.

'You have no manners,' the young woman complained.

'I didn't realize you would mind if I smoke.'

'That's not the point. You didn't offer me one.'

The poet smiled, apologized, grabbed the pack, tapped out one cigarette and offered it to the artist.

'I am not a regular. I only smoke when I am in a good mood,' she said smiling, but declined the offer. 'I only smoke menthol.'

She grabbed her bag, pulled out a book, kept it on the table, searched the bottom, retrieved a pack of menthol cigarettes and a metal lighter.

As she lit her white cigarette, the poet noticed the book. It was a novel titled *Anatomy of Life*.

The poet picked it up, flipped through a few pages and asked the artist, 'Is it any good, this novel?'

'I haven't finished reading it, as yet,' she said.

The poet handed back the novel and the artist put it in her bag.

The cigarette relaxed her. She looked around and said, 'I rather like this café. It has the charm of old times. A bit of an idyllic space is so valuable in such a crowded city. But unfortunately everything has its own lifespan.'

The poet didn't understand the last sentence. The young woman was perceptive, 'Haven't you read the notice?'

'What notice?'

The young woman pointed to the glass. A paper was pasted from the inside. The poet had missed it.

'What does it say?'

'This café is shutting down. Today is the last day. I read the notice from the street. I wanted to take a closer look at this old cafe. That's why I came in.'

The poet called the waiter and asked some questions. The waiter said the one-storey building and the land at the back had been sold. The café would be demolished soon; it could happen any day. A commercial complex was going to be built in its place.

What the waiter didn't tell them was that the owners had no intention of selling it but they had been threatened by goons hired by a local politician, who favoured a real estate developer. The nephew of the owner had also been beaten up and the local police had done nothing about the complaint due to political pressure. But the cartel had paid a reasonable price, close to the very exorbitant market rate.

The quick turn of reality numbed the poet. The future became clear to him. The street would be cordoned off by a demolition team. The noise of machines would grunt and groan through the night. Within a few hours, the plain structure would be reduced to a chaos of rubble. By early dawn, the debris would be carried away in heavy-load vehicles. With sunrise, Café Good Hope would become a memory.

'You look sad,' the young woman remarked.

'Yes,' the poet said, 'I like coming here.'

'But the old has to make way for the new,' the artist said, 'progress has to happen.'

'I understand,' the poet said, 'but progress is not all about what new things we have gained, but also about what we haven't yet lost, and have managed to preserve.'

There was an interval of silence.

The artist was taking her final look around the café when the poet proposed, 'Why don't we leave?'

10

The poet and the artist walked out of the café. At that moment, the poet described his home as a fusion of a lounge and a library, and put forward an invitation. 'My apartment is not far from here. Why don't you come along? We can drink tea, hear music and talk.'

The young woman thought for a moment. Then she smiled. 'Yeah. That would be quite nice. But maybe some other day.'

Then she said, 'I have to join a candlelight protest about violence against women. A friend of mine is the organizer.

We are also demanding judicial and police reforms. It won't take more than an hour. Why don't you come along as well? We can have dinner after that.'

The poet readily agreed and spoke of the rising trend of apolitical street protests in support of many issues, which he deemed as the positive sign of evolving consciousness.

'Yes,' the artist agreed. 'There is something in the air nowadays. It is the feeling that good can happen.'

'One has to fight for it though,' she added.

They came to a crossing. The attitude of the motorists made the cars look ill-behaved. Both of them waited at the pavement. The poet caught her arm till he felt it was safe for her to cross the street.

11

The lovers were strangers once, but they weren't strangers anymore. They spoke with the trust reserved for old friends; they were keen to tell each other the events of their lives, their thoughts, concerns and outpourings of their silence.

In other words, they sought to share their own lives, which however common or different, follow a common structure – a universal anatomy of life where seasons continue to pass with their diverse ingredients of people, places, things and circumstances; encounters with the myriad void continue to occur every day; the theme of oneself – the array of traits responding to time – keep circling within spheres of life; the centre and the periphery, the inside and the outside, continue their constant interplay; the compulsion of well-being and of being-well needs to be balanced with soul-sense; and finally the wheel, whose guardian is Time, revolves and rotates, and the cycle of birth, living and death – creation, evolution, dissolution – plays out its scheme.

As the night sneaked in silently like a deep affection, the poet and the artist mingled with the swarming faces and merged with the ocean of jostling hopes, the crowd of the metropolis.

Acknowledgements

Heartfelt gratitude to Tibor Jones South Asia Prize, Pallavi Narayan, Pranav Kumar Singh, Shiny Das, Sushmita Chatterjee and Santasree Chaudhuri, my mother, for raising me up the way she did, and for being supportive of all the crucial decisions which I have taken in my life.